THE HISTORY OF THE FUTURE

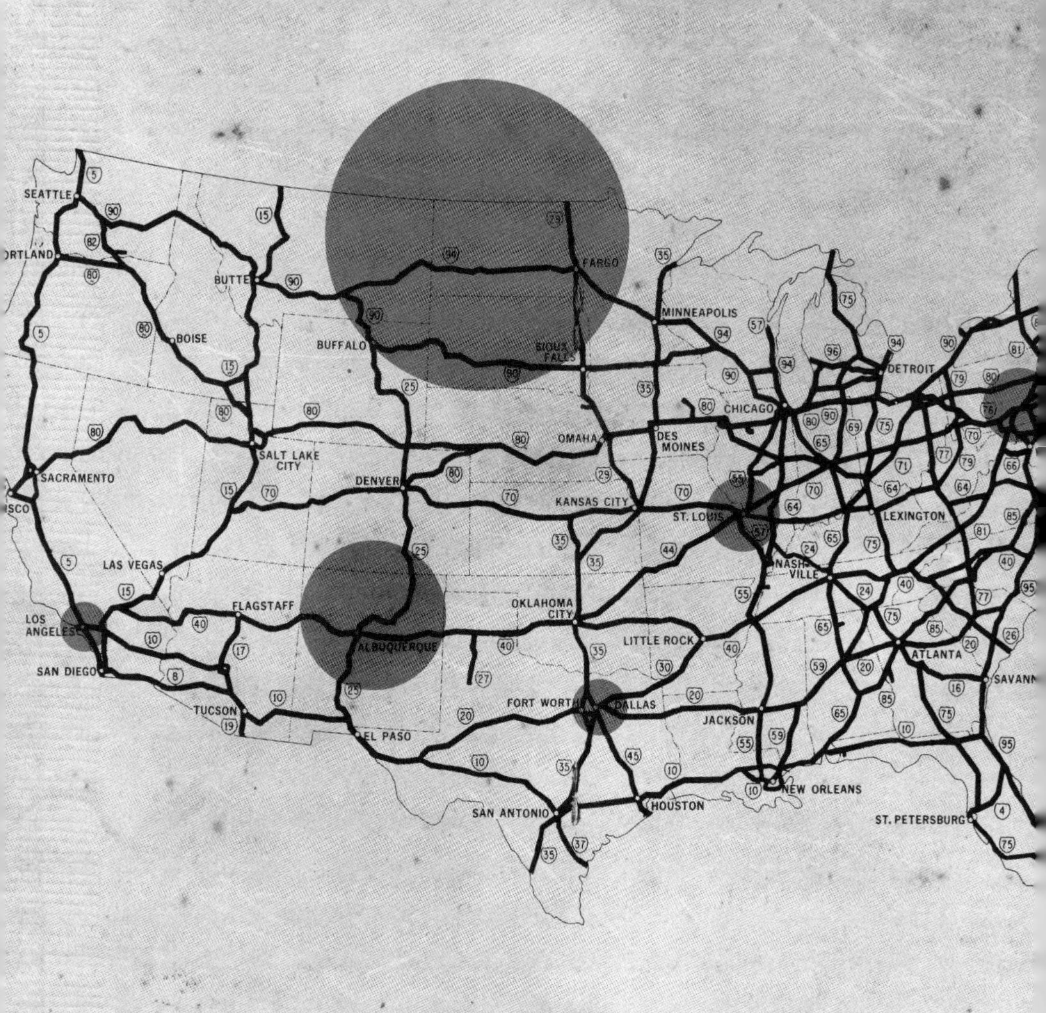
SEATTLE
ORTLAND
BUTTE
BOISE
BUFFALO
FARGO
SIOUX FALLS
MINNEAPOLIS
DETROIT
CHICAGO
DES MOINES
OMAHA
SALT LAKE CITY
SACRAMENTO
DENVER
KANSAS CITY
ST. LOUIS
LEXINGTON
SCO
LAS VEGAS
NASHVILLE
LOS ANGELES
FLAGSTAFF
OKLAHOMA CITY
LITTLE ROCK
ALBUQUERQUE
ATLANTA
SAN DIEGO
SAVANN
TUCSON
FORT WORTH
DALLAS
JACKSON
EL PASO
NEW ORLEANS
HOUSTON
SAN ANTONIO
ST. PETERSBURG

THE HISTORY OF THE FUTURE

AMERICAN ESSAYS

EDWARD MCPHERSON

COFFEE HOUSE PRESS
Minneapolis
2017

Cover design by Carlos Esparza
Book design by Rachel Holscher

Coffee House Press books are available to the trade through our primary distributor, Consortium Book Sales & Distribution, cbsd.com or (800) 283-3572. For personal orders, catalogs, or other information, write to info@coffeehousepress.org.

Coffee House Press is a nonprofit literary publishing house. Support from private foundations, corporate giving programs, government programs, and generous individuals helps make the publication of our books possible. We gratefully acknowledge their support in detail in the back of this book.

LIBRARY OF CONGRESS CATALOGING-IN-PUBLICATION DATA

Names: McPherson, Edward, 1977– author.
Title: The history of the future : American essays / Edward McPherson.
Description: Minneapolis : Coffee House Press, 2017.
Identifiers: LCCN 2016039380 | ISBN 9781566894678 (paperback)
Subjects: LCSH: United States—Description and travel—Anecdotes. | United States—Social life and customs—1971—Anecdotes. | McPherson, Edward, 1977—Anecdotes. | BISAC: HISTORY / Social History.
Classification: LCC E169.Z83 M37955 2017 | DDC 973.924—dc23
LC record available at https://lccn.loc.gov/2016039380

Acknowledgments

Earlier versions of these essays have appeared in the following journals: "Echo Patterns" and "Open Ye Gates! Swing Wide Ye Portals!" in the *Paris Review Daily*, "Lost and Found" in the *American Scholar*, "End of the Line" in *Catapult*, "Chasing the Boundary" in the *Gettysburg Review*, and "How to Survive an Atomic Bomb" in *True Story*.

PRINTED IN THE UNITED STATES OF AMERICA
24 23 22 21 20 19 18 17 1 2 3 4 5 6 7 8

Contents

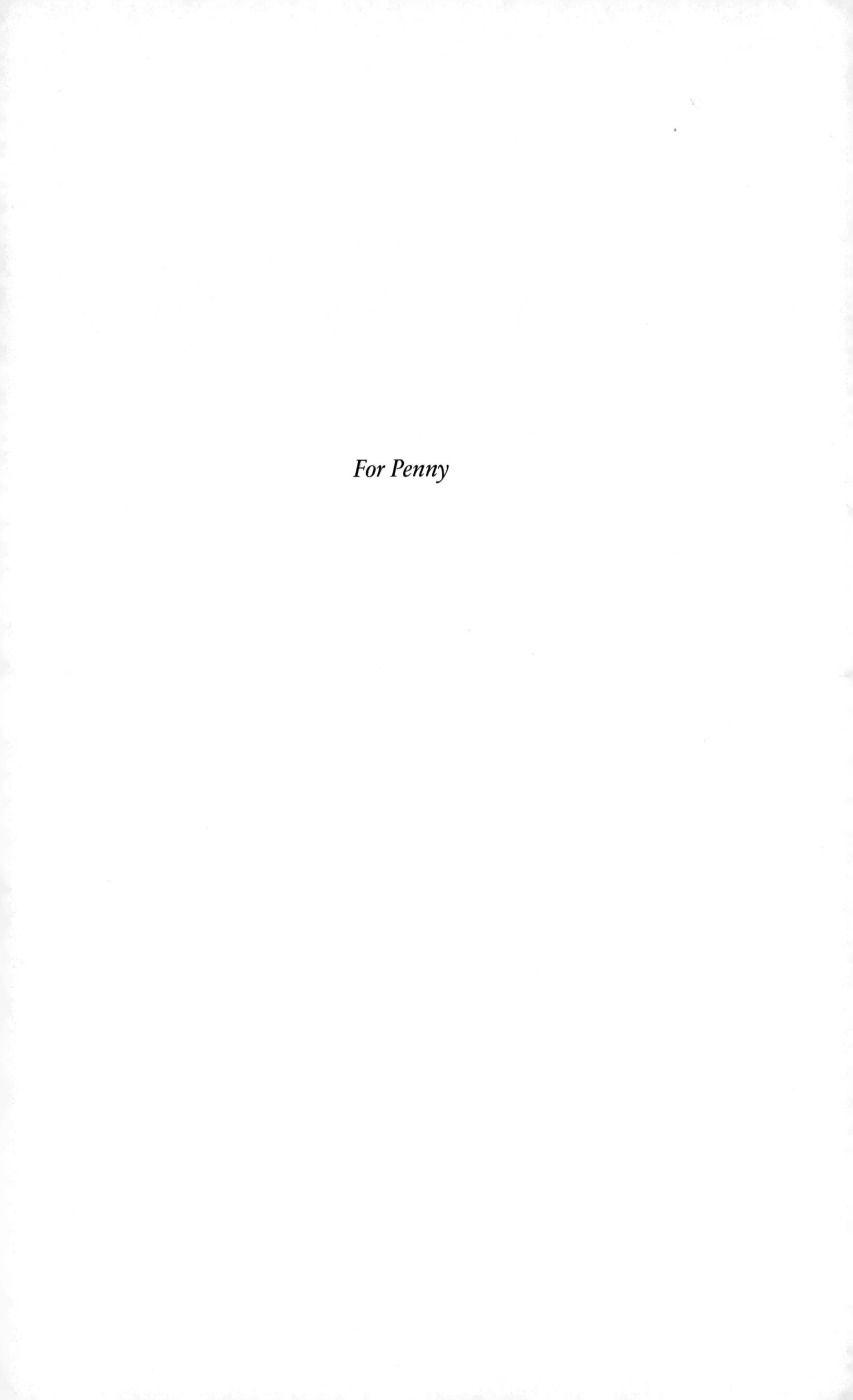

For Penny

. . . how little we can hold in mind, how everything is constantly lapsing into oblivion with every extinguished life, how the world is, as it were, draining itself, in that the history of countless places and objects which themselves have no power of memory is never heard, never described or passed on.

—W. G. SEBALD, *AUSTERLITZ*

Where you come from is gone, where you thought you were going to never was there, and where you are is no good unless you can get away from it. Where is there a place for you to be? No place.

—FLANNERY O'CONNOR, *WISE BLOOD*

THE HISTORY OF THE FUTURE

Echo Patterns

Between 318 and 271 million years ago, the ancient continental core of North America butted against what would become South America. Land folded and faulted; mountains were born. Then what would become the Gulf of Mexico opened, and inland seas washed the peaks away. It pays to remember there are mountains beneath Dallas. The tops might have eroded, but the roots remain buried deep.

Some 165 million years later—in 1841—John Neely Bryan built a shelter on a bluff and called the area Dallas.

One hundred and twenty-two years later—in 1963—John F. Kennedy was shot on that bluff, now named Dealey Plaza.

Seventeen years later—in 1980—J. R. Ewing was shot on TV.

Dallas came from nothing. Unlike surrounding areas, it was not a camp for Native Americans or prehistoric men. Dig and you find few artifacts. The Trinity River formed a boundary for ancient tribes: farmers to the east and hunters to the west. The Trinity is a true Texan; it begins and ends within the state. Its 710-mile path slices through what is now downtown Dallas, making Dallas a city on the cusp, on the boundary, in between. It wavers between being and not being. Dallas wasn't there until—suddenly—it was, called forth in the minds of white men.

John Neely Bryan, the founder of Dallas, was born in Tennessee in 1810. In 1839, he arrived at the three forks of the Trinity River with a Cherokee he called Ned and a dog he called Tubby. He was twenty-nine. He wrote his name on a piece of buckskin, affixed it to a stake, drove it into the soft ground of an eighteen-foot bluff, and went back to Arkansas. Two years later, he returned to his bluff and built a lean-to. In another two years, he was married, a union that brought five children. Dallas—as he called his claim—was on its way.

John Neely Bryan took to the bottle. In 1855, he shot a drunken man who might or might not have insulted his wife. Bryan fled to Indian Territory and hid for six years. He drifted west, looking for gold. It was a time of paranoia ("I do not write to anyone at Dallas except you, for I cannot place my confidence in any of the rest"), privation ("I was in a storm sometime since and a tree blowed down on my horse and broke him down in the line. Since which I have had to borrow"), and fantasies of revenge ("I am surprised at Colonel Stone and the other attorneys in Dallas for turning against me and I shall meet them when they least expect it and will then know the reason why they do so"). In 1861, he came back to town after hearing the man he shot had lived. His children didn't recognize him. He joined the army, but—a physical wreck—was discharged a year later. He died in a lunatic asylum in Austin in 1877.

Today, John Neely Bryan's bluff is called Dealey Plaza, after civic leader and newspaperman George Bannerman Dealey. Built in 1941, Dealey Plaza was dubbed "The Front Door of Dallas." In other words, all roads to Dallas run through Dealey Plaza.

Of course, that's not entirely true. Born and raised but no longer living in Dallas, I rarely have found myself there.

The original *Dallas* TV show aired 357 episodes from 1978 to 1991. It is a soap opera that often runs on the logic of farce: no one tells anyone else anything, and these often pointlessly kept secrets cause great misunderstanding. But from its swooping opening helicopter shot—with its swelling French horns and syncopated beat scoring a shifting triptych of skyscrapers, cattle, oil wells, football, ranchland, and Ewings—*Dallas* is surprisingly compelling television: a bunch of beautiful people making disastrous choices at a spectacular rate. My wife remembers watching the show with her family growing up in Missouri. Rewatching the series now, when a shirtless Patrick Duffy surfaces dripping from a pool, she says, "I think that's the chest I imprinted on." There passes an uncomfortable moment, but, in a way, she's right—the rest of the country imprinted on *Dallas*, too.

While locals derided the show's inaccuracies (all those California cowboys stepping gingerly about in their ropers), the original series was rooted in a sense of place. It was a dream of a certain expression of wealth, taste, and desire that always drove back to one thing—family. How quaint all those communal meals seem now, with three generations meeting at the table for breakfast *and* dinner (making sure to phone when they plan to be out all night on their nefarious errands). The show's constant refrain, "You're a *Ewing*," meant, "Act like one." Southfork—such small potatoes to the McMansions of today—was a prison. The series was a journey into the claustrophobic dark heart of familial dysfunction. The central, often paranoid, concern: *us* against *them*. And that is how the show got Dallas right.

Former Dallas Mayor R. L. Thornton, on December 4, 1963, less than two weeks after Kennedy was gunned down: "I've heard people talking about erecting a monument in their sadness. For my part, I don't want anything to remind me that a president was killed on the streets of Dallas. I want to forget."

According to the official *Dallas* fansite: "It wasn't until its 1979–1980 season cliffhanger when J. R. was shot that the show was catapulted into a worldwide phenomenon." In other words, Dallas wasn't *Dallas* until it made its rhyme with history, until it acknowledged its own past.

Then again, perhaps *Dallas* wasn't about Dallas at all. There's something suspiciously translatable about a show that—in its heyday—aired in ninety-six countries in fifty-five languages.

December 1846: The first civil suit in Dallas goes to court, in which Mrs. Charlotte M. Dalton sues Joseph B. Dalton for divorce. The divorce is granted. Within a few hours, Mrs. Charlotte M. Dalton finds herself back in the legal register, this time for wedding Henderson Crouch, the foreman of the previous jury. Certainly this is not the first instance of intrigue and infidelity in the new city, but it does read something like a proto-*Dallas* plot.

In the middle of downtown, not far from Dealey Plaza, stands a cedar log cabin. Like most Dallasites, I grew up thinking it was Bryan's original structure, but it turns out it's probably a replica. (Theories and counter-theories abound.) They say John Neely Bryan named the city after his friend, Dallas, but who that friend was has been lost to history.

In 1855, some two hundred French, Swiss, and Belgians—some on horses, some on foot, some in wooden shoes—made the 250-mile trek from Houston, settling just west of Dallas in a utopian community they called "La Reunion." The settlement was a cooperative founded on the social ideals of French philosopher Charles Fourier. Women were equal to men and could vote. The usual entropy—combined with substandard housing, a severe winter, a spring drought, summer grasshoppers, and a crop of wheat grown without consideration that there was no one there to buy it—undid the community's best-laid plans. By 1860, 160 members of the colony had defected to Dallas. Thus the first piano entered the city, which also gained from a pool of pastry chefs, brewers, dancing masters, artisans, jewelers, tailors, physicians, naturalists, and the like. The seeds of Dallas as a cultural hub were planted. La Reunion ended without formal dissolution—it simply disappeared, save for a small cemetery, once called Fishtrap, which—some seventy years later—would become the resting place of outlaw Bonnie Parker. Today the cemetery remains overlooked, overtaken by a different kind of entropy: weeds, litter, a chain-link fence.

Reunion Tower—named in honor of the colony—rises roughly three miles to the east, standing tall like a late-1970s microphone.

Dallas is: Big Tex, the Cotton Bowl, the Dr. Pepper clock, Baby Doe's Matchless Mine (as seen from I-35), El Fenix (the downtown original), Reunion Tower, Fair Park (home of the state fair as well as the world's largest collection of Art Deco buildings), Texas Stadium (RIP), Old Red Courthouse, Highland Park Village (the country's first planned shopping center), NorthPark mall (with its modern art and holiday penguins), Love Field, the Cathedral Santuario de Guadalupe, Neiman Marcus, Bank of America Plaza (a.k.a. the "Green Building," built in 1985 and outlined with two miles of green argon lights), the aluminum-clad Republic National Bank building (its rocket-like spire stretching to the sky), First Interstate Tower (its giant slanting curtain walls once home to Ewing Oil), Texas Commerce Tower (its curved glass top pierced by a seven-story hole, a passage made narrower than originally planned for fear that some Dallas daredevil would try to fly a plane through it), the beaux-arts Magnolia Building (boasting, in 1934, the largest rotating sign in the world, a six-thousand-pound red neon Pegasus, the logo of Mobil Oil, now ExxonMobil, headquartered thirteen miles down the road), among others.

Some of these skyscrapers were to be built in pairs, but even in Dallas developers can't forestall a crash in real estate or a bust in oil. So the buildings stand, twinless ghosts on the skyline.

John Steinbeck on Texas: "Few people dare to inspect it for fear of losing their bearings in mystery or paradox." Dallas, in particular, is a city that resists narrative. On my first day in New York City as a young, starry-eyed writer, I recall sitting in the corner office of the fiction editor of the *New Yorker* and being asked, "Why is there no great fiction from Dallas?" My bookshelf brims with titles like *The Lusty Texans of Dallas; Texas, My Texas; The Big Rich; Diaper Days of Dallas; The Cultural Studies Reader; Dallas: A History of "Big D"; Dallas: The Making of a Modern City; In a Narrow Grave.*

A list of movies shot in Dallas: *State Fair, Bonnie and Clyde, The Killer Shrews, Mars Needs Women, Benji, Logan's Run, True Stories, Talk Radio, The Trip to Bountiful, Born on the Fourth of July, Bottle Rocket, Batman & Robin, Office Space, Problem Child, The Apostle, Armageddon.*

A film not shot in Dallas: *Debbie Does Dallas.*

In *RoboCop,* Paul Verhoeven's 1987 dystopian warning about the perils of the corporate police state, the bloodthirsty private-sector executive is hurled out of the dramatically canted windows of Dallas City Hall—here representing the headquarters of the evil corporate monolith OCP, which is bent on taking control of the city. Villains and cops chase each other through the Dallas streets; at times, Reunion Tower can be seen in the background. Of course, in the film, this calamitous, nearly postapocalyptic vision does not unfold in Dallas, but Detroit.

Dallas City Hall, built in 1978 by modern architect I. M. Pei, is an upside-down pyramid sunk into the ground. The project was part of the city's "Goals for Dallas" effort to rebrand itself after the Kennedy assassination. Pei is quoted as saying, "When you do a city hall, it has to convey an image of the people, and this had to represent the people of Dallas. . . . The people I met—rich and poor, powerful and not so powerful—were all very proud of their city. They felt that Dallas was the greatest city there was, and I could not disappoint them."

The effect of the showstopping cantilevered façade was patently obvious to the makers of *RoboCop:* the blocky concrete building—which slopes to be wider at its top than at its base—is an architectural bully, looming overhead, dwarfing the individual and threatening to crush all who dare enter the halls of civic duty.

Dallas is geographically cursed: it bears no significant lakes, mountains, waterfalls, or waterways. (In time, settlers would learn the Trinity River was not navigable.) Nineteenth-century advertisements for the new settlement celebrated a climate that "for health and pleasure is not surpassed by any in the world, and in this respect may be termed the Italy of America." It was a clear case of wishful thinking. The city lies a little above thirty-two degrees north, meaning it more or less shares a latitude line with Charleston, Marrakesh, Tripoli, Nagasaki, and San Diego. But its weather is its own. Dallas winters can be cooler than you might imagine; summers are unbearable. The 1950s saw a milestone for the city: the rise of the residential air conditioner.

Dallas, an Incomplete Timeline

July 1872: The north–south Houston and Texas Central railroad arrives in Dallas to tremendous fanfare.

1873: The east–west Texas and Pacific railroad crosses the H&TC railroad—at Dallas. The city becomes a boomtown.

Two years earlier: The sneaky delegates of Dallas insert a seemingly innocuous rider into a bill in Austin that gives the Texas and Pacific Railroad company—now laying track along the 32nd Parallel, about fifty miles south of Dallas—statewide support and general leeway to plot its transcontinental course across Texas, provided that the line run within a mile of the little-known marker, Browder's Springs—which was later revealed to be just south of downtown Dallas.

1873: The first theater in Dallas—really a second-floor auditorium without adequate space or dressing rooms—opens on Main Street, less than a mile from what is now a bustling cultural district.

1881: First telephone.

1885: First issue of the *Dallas Morning News,* circulation five thousand.

1897: First projection of Edison's Vitascope, a year after its New York City premiere. Moviegoers gather at the opera house to witness a hanging, a lynching, dueling Mexicans, a fiery rescue, and scenes of Niagara Falls.

1899: The first automobile in the state of Texas is driven into Dallas by one-legged owner and local railroad magnate Ned Green, who piloted the two-cylinder two-seater from Terrell, where it came off a train. The thirty-mile trip takes more than five hours, Green terrifying horses and livestock along the way with speeds of up to twenty-five miles per hour.

1905: Businessmen band together to promote Dallas as "the City of Splendid Realities."

1907: The first skyscraper appears in Dallas, towering at fifteen stories. Neiman Marcus opens.

1914: Dallas becomes the smallest city in the country to get a Federal Reserve bank.

October 24, 1923: Ku Klux Klan Day at the Texas State Fair. Twenty-five thousand fairgoers see 5,631 men sworn in to the Dallas chapter of the KKK, the largest in the country at some thirteen thousand members strong. The *Dallas Morning News*—and many of the city's old guard—is against the Klan, which had undergone a surprising resurgence after being essentially absent since the last century. The paper remains unwavering in its opposition, and public support for the Klan dwindles. By 1926, the Dallas chapter numbers thirteen hundred. Three years later, it closes its doors.

1925: For the second time in its history, Dallas hosts the national reunion of the Confederate Veterans of the Civil War, despite that Dallas is not truly "Southern"—but Texan. The city first hosted the group in 1902, when the entire town devoted itself to the celebration. In 1925, the veterans are treated with care, but to many Dallasites it is simply another of many conventions held in the city.

1935: Gertrude Stein lands in Dallas and is greeted by eight thousand fans, who are under the impression her plane is Clark Gable's. At her lecture, a reporter for the local paper finds her "perplexingly clear."

1936: Dallas hosts the Texas Centennial Exposition, which runs for nearly six months and sees six million visitors. Two years earlier, in a surprising display of ahistorical chutzpah, Dallas won the bid to hold the convention, beating out San Antonio, Houston, and other cities that actually existed when Texas was founded.

1957: Maria Callas makes her debut in Dallas. *Time* magazine crows, "As of today, Dallas is on the map as an opera town along with New York, San Francisco, and Chicago." Next came *Newsweek:* "For a couple of nights running last week, Dallas, Texas, was the operatic capital of the U.S."

1957 (again): Let us stop here, six years before the Kennedy assassination, and compare the crime statistics from Dallas to those from Atlanta, a city of a similar size. In 1957, Atlanta had 17 percent more murders than Dallas, 50 percent more aggressive assaults, and more than twice the number of thefts over fifty dollars.

Dallas, an Incomplete Timeline

Late 1977: Fledgling TV writer David Jacobs presents to CBS the idea for a glitzy drama called "Untitled Linda Evans Project." Originally inspired by the work of Swedish auteur Ingmar Bergman, the show's setting is changed to Texas. Linda Evans is not involved.

On December 10, Jacobs submits a pilot for the show, now called *Dallas*, despite having never visited the city. Even he objected to the title—pointing out Houston was oil, Dallas was banking—but a network executive told him, "Dallas is more commercial." Five days later, the network orders five episodes to complete the miniseries.

April 2, 1978: Bobby Ewing brings his new wife, Pamela, to Southfork to meet his family. She is the daughter of his father's oldest enemy. Bobby is Romeo; she is Juliet. Pamela tells him, "Your folks are gonna throw me off the ranch." She is not thrown off the ranch, but—in the first episode—she is accused of being a spy and subject to attempts to bribe, seduce, and frame her.

And so we meet the major players: Jock and Miss Ellie, the family patriarch and matriarch whose combined fortunes (Jock's oil, Miss Ellie's land and cattle) have made Southfork what it is. J. R., the slick and scheming firstborn up to his ten-gallon hat in shady oil deals. His wife, Sue Ellen, a boozy former Miss Texas. Bobby, the youngest—and best—brother, trying to prove himself in the family business. His wife, Pamela, the daughter of Jock Ewing's former business partner, a wildcatter named Digger Barnes, who claims he was swindled out of a fortune. Pamela's half brother, Cliff Barnes, a lawyer on the make who wants to bring down the Ewings. Lucy Ewing, the daughter of the prodigal middle brother, Gary, who couldn't handle the pressures of being a Ewing and ran off, leaving Jock and Miss Ellie to raise Lucy, who—as a high schooler—is having an affair with a much older ranch hand who later will turn out to be her (illegitimate) uncle.

Typical goings-on around Southfork: sexual blackmail, disco dancing, the threat of rape, gunplay, adultery, sex with a secretary, secret stolen files, betrayal, the ruination of a senator, a hostage situation, a rendition of "People" performed at gunpoint by a former beauty queen in a bathing suit, a pregnancy (Pam's), a push out of a hayloft by a drunken J. R. (Pam's), a miscarriage (Pam's, after falling from the loft), the return and subsequent disappearance of a prodigal son, open-heart surgery, revenge-seeking hicks, a hurricane, secret bigamy, and a black-market baby. And that's only in the first eleven episodes.

March 21, 1980: Alone in his office late at night, J. R. receives a phone call. He answers, but the caller hangs up. J. R. gets coffee and stares out the window. The camera follows the point of view of a mysterious intruder creeping into the office. J. R. hears a noise: "Who's there?" He goes to investigate. As he steps through the door, he is shot twice and falls to the floor. Credits roll. End of season.

August 11, 1980: J. R. appears on the cover of *Time* magazine under the word "Whodunit?" Conspiracy theories abound; potential culprits include his brother, his wife, his mother, his brother-in-law, his wife's lover, his wife's sister (who is also his ex-mistress), an angry henchman, and a crossed business partner, among others. Las Vegas even eventually puts the odds at twenty to one that the shooting was somehow a suicide.

November 20, 1980: The *New York Times* reports, "Southfork now rivals Dealey Plaza and the Texas School Book Depository as Dallas's most popular tourist attraction."

November 21, 1980: Eighty-three million Americans watch J. R.'s wife, Sue Ellen, learn through hypnosis that she didn't shoot J. R.—it was her little sister, Kristin! The revelation plays on 76 percent of the televisions turned on *in the country.* This viewership is dwarfed abroad, where three hundred million people witness the mystery's solution, including the members of Turkish parliament, who end business early in order to make it home to watch.

December 11, 1981: Bobby buys Kristin's son, Christopher, off a drug dealer and brings him home to Southfork. Christopher becomes a pawn in the battle between Bobby and J. R., before eventually being adopted by Bobby. He will fire a gun at J. R.'s son, John Ross (his cousin), before he turns six.

January 1, 1982: Miss Ellie receives the news that her husband, Jock, has died in a helicopter crash. With their father gone, the brothers' feud flares up unchecked.

May 17, 1985: Bobby Ewing pushes Pam—now his ex-wife—out of the path of a speeding car and is thrown over its roof. The driver, Pam's crazy half sister who is disguised by a wig, dies instantly. Bobby dies in the hospital.

December 1985: *Het Geval Dallas,* a work of ethnographic research done on viewers in Holland, is published in English under the title *Watching Dallas.* The author, Ien Ang, discovered that those who hated the show *really* hated it—with a fury that exceeded the logic of their objections (of sexism, materialism, sensationalism, etc.). Other viewers of the show used irony to distance themselves from it—and thus allowed themselves to enjoy it. One finding: "While liking *Dallas* ironically leads to euphoria and merriment, as we have seen, disliking *Dallas* is accompanied by anger and annoyance. . . . Hence those who dislike *Dallas* run the risk of a conflict of feelings if, *in spite of this*, they cannot escape its seduction."

May 16, 1986: Pam remarries. Sue Ellen, out of rehab and reconciled with J. R., is blown to bits by an explosive briefcase left in the offices of Ewing Oil. The morning after her wedding, Pam wakes to find Bobby in her bathroom, taking a shower. He says, "Good morning." The past year has been a dream, thus nullifying her remarriage and the deaths of Bobby, Sue Ellen, her half sister, and her brother's wife, among others.

May 3, 1991: In the series's final episode, "Conundrum," a suicidal J. R. is visited by a demon that treats him to a vision of what life would be like if he had never been born. Some people fare better, others worse. In the end, J. R. puts a gun to his head. The devil shouts, "Do it!" A shot rings out. Bobby bursts into his brother's room and says, "Oh my God." The frame freezes on Bobby's face. The credits roll.

In the aftermath of the assassination, 80 percent of the country blamed the people of Dallas for Kennedy's death. Dallasites found themselves disconnected by long-distance operators and—when they traveled—harassed in gas stations, restaurants, and cabs. Two days after the assassination—and ninety minutes after Oswald was shot—the Dallas Cowboys took the field in Cleveland to the jeers, "Kennedy killers." (A lineman remembers, "They called us everything—murderers. Like we had something to do with it. We didn't get over it for weeks.")

A card left in Dealey Plaza the day after the shooting pleaded, "God forgive us all."

Dallas's psyche was shattered. There was a statistical increase in murders, suicides, and coronary deaths. Dallasites felt profound embarrassment and shame. But "City of Hate" did not fit into Dallas's narrative of optimistic exceptionalism, and the city went to work reinventing itself thanks to a new mayor culled from Texas Instruments, a tech company with its eyes on the bright, shining future, not the oily past. Within seven years of the assassination, *Look* magazine labeled Dallas an "All-American City." Two years later, the Dallas Cowboys won their first Super Bowl, fast on their way to becoming "America's Team." Six years later, *Dallas* hit the air.

That is not to say that, at times, the city can't be defensive, resentful, and forgetful. In 1988, a study revealed some 70 percent of Dallasites still felt that non-Dallasites held them to blame. After the assassination, Texas man of letters Larry McMurtry wrote, "Wealth, violence, and poverty are common throughout Texas, and why the combination should be scarier in Dallas than elsewhere I don't know. But it is: no place in Texas is quite so tense and tight . . . Dallas is a city of underground men: the violence there lies deeper, and is under greater pressure."

Men and women who can be found underground in Dallas: Clyde Barrow (outlaw), Bonnie Parker (outlaw, buried in a separate cemetery than Barrow), Tom Landry (Cowboys coach), Mickey Mantle (Yankee slugger), and "Mary Kay" Ash (cosmetics mogul).

The business of Dallas is beauty. An old saying: "Texas is all right for men and dogs, but hell on women and horses." A film clip from 1935: Depression-era women seek jobs as models at the upcoming Texas Centennial Exposition. A line of women, in bathing suits, is examined by suited men with clipboards who turn them and grab their legs and shoulders before squeezing them into life-sized cutouts of the ideal female form. One woman jokingly hoists her leg to place a foot against the cutout and leans back to yank out her friend, whose posterior is presumably stuck in the frame.

The business of Dallas is fashion. Since its humble beginnings in 1907, Neiman Marcus has given the people of Dallas a claim to taste and refinement that exceeds what might be expected of a city on the north Texas plains. Merchandising genius Stanley Marcus turned the department store founded by his aunt, uncle, and father into an international luxury brand—along the way personally selling more than ten million dollars in furs to his customers, as well as inaugural gowns to at least two first ladies. Neiman Marcus—known as "Neiman's" or "the Store"—offered luxury and rarity, but it was the imprimatur of good taste that ensured its success. (Gone were the anxieties of the Texan traveling abroad or the sudden possessor of "new money"—by shopping at Neiman's, they were guaranteed to be safe, clad in a certain kind of style.) The store also embraced the notion that the good life is part drama, part dream. Since the 1960s, its Christmas catalog has offered his-and-her submarines, airplanes, mummy cases, robots, windmills, and the like; the 2012 holiday gifts include a walk-on role in Broadway's *Annie* ($30,000), a water-propelled jetpack ($99,500), and a McLaren 12C Spider convertible "supercar" ($354,000), all twelve of which sold out in under two hours.

Dallas has always had similar priorities. The first shot of the entire series is a close-up of the Mercedes symbol on a red 450 SL with black leather seats. The license plate reads "Ewing 4." Bobby and Pam, just married, are heading to Southfork in style.

The business of Dallas has always been business. A town of crude, cotton, fashion, and banking. Pressure, time, and decay make oil—but the irony, of course, is that there is no oil beneath Dallas. The city is surrounded by wells in nearly every direction, but at the center it is empty, like a donut.

A colleague once told me a story. She grew up during the Cold War in a city on the Romanian border. Under Communism, there was never enough food. It was illegal for citizens to start farming on the property of the state—even if the farms were small vegetable gardens, and even if that property wasn't in use—but there was some neglected land on the banks of the river that—because it was in the floodplain and often underwater—didn't exist on any official maps. In secret, her parents, along with four or five families in her apartment building, began growing food on that land. Soon, it became an organized social hub. For years, they would go to this commune, while the authorities turned a blind eye, content, perhaps, that the letter of the law was being observed.

What did these people behind the iron curtain call this communal space—of sustenance, of freedom, of dreams—that did and did not exist?

They called it *Dallas.*

Dallas is an underdog. Landlocked, not blessed by a navigable waterway, Dallas made itself into a transportation hub by sheer will.

Charles Lindbergh, at a banquet in Dallas in September of 1927, told the city, "Keep your airport—it will place you among the commercial leaders of the world."

A midcentury scene: an aviation company is considering moving to Dallas. The president of the company is heard saying the runways at Love Field aren't long enough by two thousand feet. Three hours and forty minutes pass. The phone rings; it's the city council. Thanks to an emergency bond measure, crews will begin lengthening the runways on Monday.

But Dallas eventually outgrew Love Field and built itself a bigger airport, Dallas–Fort Worth International, which opened in 1974 and now sprawls some twenty-seven square miles, making the airport the third largest (in terms of size) and fourth busiest (in terms of takeoffs and landings) in the world. You could fit JFK, LAX, and O'Hare within its boundaries and still have room to park. Thanks to DFW, the people of Dallas are within four hours of every major U.S. city in the lower forty-eight. Not that all Dallasites feel the need to travel—what they don't have, they build.

Despite a real-estate boom and bust, dark days from the mid-1980s to the mid-1990s when oil prices plummeted and banks collapsed, Dallas is back riding tall in the saddle. During the recent recession, it ranked among the most robust economies in the country. The city built a new bridge across the Trinity River and opened a massive, newly revitalized arts district, featuring, among other buildings, a sculpture garden designed by Renzo Piano, an opera house designed by Foster + Partners, and a performing-arts space designed by Rem Koolhaas.

On June 13, 2012, *Dallas* went back on the air, picked up by the cable channel TNT and updated to include hackers, sex tapes, smartphones, cloud drives, and a real-life website where you could read the characters' tweets (and tweet back at them). There might have been more French horns, but it was still your parents' *Dallas:* the show was not a remake or a reboot, but a continuation—with the writers attempting to violate nothing in the convoluted backstory of the series. (How would you reboot these characters, who seem so real?) J. R., Bobby, Sue Ellen, and Cliff were back, supported by the next generation of Ewings and their mates. The first season—which averaged 5.3 million viewers and ranked as the top new adult cable drama—trafficked in the familiar: a Ewing barbecue, a health scare (this time Bobby's, not Jock's), secret identities, shifting loyalties, star-crossed love, and rapid-fire cliffhangers (the new show-runner has said she wants two to four *an episode*). Meanwhile, three to four hundred fans visit Southfork *a day.*

How does J. R. play today? Since 1991, the country has seen its share of modern Machiavellis—good old boys for whom truth is just a malleable means for power—but none can top J.R., the Lone Star scoundrel, the frontier incarnate one hundred years later in a trim power suit. (For those raising a hand in protest, George W. was never really a rancher; cowboys don't attend boarding school and the Ivy League—and they certainly don't summer in Maine.)

Dallas is J.R.—he was the only cast member to appear in all 357 episodes of the original series—and, up until the actor's death on November 23, 2012, J.R. was Larry Hagman. Hagman was and was not a real cowboy. He was born in Fort Worth and graduated from Weatherford High School. But he also spent time in California and New York; and, of course, his mother was actress Mary Martin—a.k.a. Peter Pan. His first big gig was playing Major Anthony Nelson in *I Dream of Jeannie.* He once lived on a ranch called "Heaven" in California, but years of hard living caught up to him. In 1995, he had a liver transplant, and days before shooting began on the new *Dallas,* he found out he had throat cancer, meaning that while Bobby Ewing was battling cancer in season one of the new show, it was J.R. who was undergoing real-life chemotherapy.

At the start of production on season two, Hagman was reported to be cancer-free and back in good health. His hair was thinning but his eyebrows were thick, sprouting out of his head and shooting skyward like the legs of a locust or, if you prefer, the horns of a Longhorn. Or devil.

The eyebrows were a subject of great discussion on and off the set. But they were not up for debate. Hagman insisted they not be touched—J.R. does not manscape—but, given the demands of HD, the head of makeup was allowed to apply a little product. But should she want to trim a single hair, she had to ask permission.

At the time, Hagman ran an active charitable foundation that operated under the slogan "Evil Does Good."

Is it too much to say Dallas never stopped being *Dallas*? Yes, the show is a soap opera, but it seems significant that the show and its viewers would stand for a whole season, from 1985 to 1986, that was a dream—and thereby erased a number of unspeakable tragedies.

And, sure, people grow up and grow old, get jobs and get married, live busy, honorable, and pleasant lives in Dallas, as they do in any other place. It is, of course, just another city. But there exists another Dallas. And for many—perhaps most—people, the actual Dallas has been overtaken by the other Dallas, that Dallas of the mind (which some might call *Dallas*). I'm not downplaying the death of a real president, or ignorant of the political and social consequences of assassination and alienation—but I want to be clear that the nation's experience of that death (and that town) is understood mainly in symbolic or abstract terms. Squint your eyes and JFK becomes J.R., who was shot everywhere and nowhere. I know it sounds profane. My mom hates the fact that most of the country only knows Dallas for an assassination and a TV show; perhaps she will not like this essay.

For her sake, among other reasons, I decided to visit Dallas once again.

> *Have you ever seen Dallas from a DC-9 at night?*
> *Dallas is a jewel, Dallas is a beautiful sight.*
> *And Dallas is a jungle, but Dallas gives a beautiful light.*
>
> —JIMMIE DALE GILMORE, FROM THE SONG "DALLAS"

From a Boeing 737 on a sparkling fall day, Dallas looks like a patchwork of mottled greens and browns, the ground more rich and loamy than withered and sere, as if the coming winter were just nature's way of winking. The lakes are murky, the land billiard-table flat, laced with former wagon trails that have become thoroughfares. Approaching the city, cloned suburban houses sprout in rows that curl and stretch with predetermined whimsy, the pools, tennis courts, and golf courses popping up at neat intervals. Divided expressways thread the map, the roads laden with cars, pickups, motorcycles, and semis all going, going, going, even on a Sunday, even on a *football* Sunday.

I am flying into Love Field, an airport that has served Dallas since 1917, when the army named the flying field after First Lieutenant Moss Lee Love, who crashed and died in his Type C Wright pusher biplane four years earlier. Kennedy landed at Love Field at 11:37 a.m. on November 22, 1963. It is a Texas State Historical site. I am flying into history.

Twice the plane lowers its right wing as we approach downtown, as if in deference to that storied skyline. A spindly white bridge—designed by Spanish "starchitect" Santiago Calatrava—swoons in the distance, a cobweb of cables crossing the river. I am not approaching from that most famous angle, fast and low up the Trinity River bottoms, but from the east, and so we fly over LBJ freeway, then the State Fair grounds, with the Cotton Bowl and the ruins of Big Tex, the giant motorized cowboy who, with his seventy-five-gallon hat, pointed the way to Fair Park until he burned down nine days before my flight. (In the aftermath of the fire, the mayor of Dallas tweeted: "Dallas is about Big Things and #BigTex was symbolic of that. We will rebuild Big Tex bigger and better for the 21st Century." Of course Big Tex himself was a rebuilding/rebranding job. The fifty-two-foot-tall cowboy began life in 1949 as a colossal Santa Claus, before being installed—three years later—at the fairgrounds. From one dream to the next.)

We skirt the Park Cities neighborhood, keeping downtown to the south. The grass is still growing; the trees bear leafy green tops. And then the wheels hit, and the man in front of me—black, in his late thirties, a diamond stud in one ear and a Gucci carry-on in his lap—says loudly to no one in particular, "Home sweet home." Meanwhile, in a stadium across town, the Dallas Cowboys kick off against the New York Giants.

We stagger onto the bright jetway, which leaks sun and refreshing sixty-three-degree air. Below, baggage handlers work in shorts and wraparound shades. At the gate, a frazzled woman is complaining to a listless security agent that there is no place to smoke in the terminal.

Behind the woman a flat screen carries the Cowboys game. In fact, as far as the eye can see there are TVs showing the game. No sign of CNN or the Weather Channel here, despite the impending landfall of "Superstorm Sandy" that is wrecking travel for roughly one-third of the country in the Northeast. Last season, the Giants knocked the Cowboys out of a play-off spot. This is the first home matchup since then. Revenge weighs heavily on the city's mind.

Three months ago, in July, the team's owner, Jerry Jones, exhorted fans at a rally: "Y'all should come to that stadium and watch us beat the Giants' ass." This week, management sent an email to season-ticket holders asking them to be loud, particularly on their opponent's third downs, so that the shame of the last home game—when a bunch of visiting Bears fans out-whooped and out-hollered and generally embarrassed the hometown supporters—doesn't happen again.

At halftime, the Cowboys trail 23 to 10.

They come back 23 to 17 with nine minutes to go in the third.

Five minutes later, Dallas is up one point, 24 to 23.

The Cowboys haven't ever beaten the Giants in their new stadium.

Two field goals later—with 3:31 left in the game—New York is ahead, 29 to 24.

The game is a nail-biter. Deep in the Giants' territory, with only a little more than a minute on the clock, the Cowboys throw an interception on fourth down after failing to convert a second-and-one. The game has taken on an air of fiction: the Cowboys, a troubled, disappointing team this season, won't lie down and die—week after week, they find inventive ways to lure even the most wary fans into believing again, only to fail them. It's as if the sportswriters in the sky were coming up with more and more outlandish plots to tug at a city's heartstrings.

A friend of mine shouts, "This team maximizes pain!"

The Cowboys gets the ball back with forty-four seconds.

My friend covers his face and says, "I can't believe I still care."

A touchdown would win the game.

Twenty-five seconds: they're inside the Giants' territory.

"They always do this to me. They always do it." My friend rocks on his heels. "This is the entire season."

Then, with sixteen seconds on the clock, a miracle: a high, unexpected Hail Mary into double coverage—pure prayer, what we used to call a "huck" on the playground—is caught by a receiver leaping into the back of the end zone. The stadium levitates. The Giants' coach looks like he's suffered a coronary. My friend bolts out of the room, shouting gibberish.

Then: there's a chance the receiver landed out of bounds.

Commercial break while the officials review it.

Cameras dissect the play from every possible angle. No one can agree on what he sees. The receiver's hand seems to graze the ground an instant before the rest of his body crashes into the end zone. Where do his fingers land—out or in? From the air, his fingers stretch out, out, out—then his palm, elbow, and body land in bounds. But it is no matter. The officials make their ruling: out of bounds by a finger. By a millimeter. No touchdown.

My friend shouts, "Conspiracy!"

Outside the window, in the darkening twilight, you can hear the city groan.

But it's still not over. Ten seconds left: a pass to the sideline, a handful of yards.

The Cowboys throw an incomplete pass across the middle.

New York calls a time-out.

One second left. The last last chance.

The Cowboys throw the ball out of the end zone.

Hope and heartbreak.

The Giants win and will now have to fly back into Hurricane Sandy.

The mythic allure of the Dallas Cowboys—a subject worthy of many of the countless books it generates—is too great a phenomenon to fully unpack here. Founded in 1960 by the sons of oil barons, the team is steeped in tradition and oil. As Dallas searched for something to cheer about in the years after Kennedy, the Cowboys became a lone star rising above the new Texas: "America's Team," as they became known, a symbol of the biggest and best in the state. The Cowboys have sold out every home game since 1991, even in their new billion-dollar stadium—the largest in the league—which crammed in a record 105,121 fans, some standing, some sitting, for its opening in 2009 (another game in which fans watched the Cowboys fall in the final seconds to the Giants). *Forbes* lists the team as the world's second most valuable sports franchise (after England's Manchester United soccer club), and even sixteen years after the Cowboys' last Super Bowl victory, the team matters more than its wins and losses. The franchise is worth $2.1 billion, thanks to the aggressive—and some might say cutthroat—deal-making of oilman Jerry Jones. (The team makes nearly $20 million more off its stadium in sponsorship deals than any other NFL team; its operating income is $108 million greater than the next runner-up's, a sum that exceeds the *entire take* of the NBA or the NHL.) But money only tells part of the story. The Cowboys remain one of the premiere characters—sometimes heroes, sometimes villains—of the NFL in a way that the Cleveland Browns, say, will never be. Its cheerleaders have their own reality TV show, now going on seven seasons. Currently, the Cowboys are a dreadful team; Jones is regarded as a vainglorious meddler, a plastic-surgery-enhanced and spectacle-driven egomaniac, ever appearing on the sidelines to mismanage his fractured, overpriced team. But at this point, the image has supplanted the reality. This year, an ESPN poll found the Cowboys to be the most popular team—yet again—with nearly 9 percent of fans nationwide rooting for them each week.

I'm in a third-floor deluxe suite with a sitting room, two bathrooms, four-hundred-thread-count sheets, and an oblique view onto a pool—all of which, according to the rate on the door, goes for somewhere between $900 and $1,000 a night. Of course I'm not paying that. I'm staying with one of the writers of the revamped *Dallas,* a childhood friend who has flown in from L.A. and has occupied this room for the past three weeks. There is a television embedded into the bathroom mirror.

Across the room, at a desk that contains stacks of scripts—pink pages, blue pages, a day's worth of scenes shrunk down to pocket-size—along with a telephone, a laptop, a paper coffee cup, a plastic lighter, and a fifth of Black Label, sits my friend. He is writing, putting words into the mouths of characters who tomorrow will speak them, whereby—post-production—they will form another gospel in the Holy Writ of *Dallas,* to be debated and discussed on Twitter, fansites, glossy magazines, and the like. I'm on the couch, writing about him.

Exterior, late October 2012: a nondescript south Dallas studio, not far from John Neely Bryan's cabin, where another Dallas—the original Dallas—was born. Looming above, a massive industrial block, bombed out, tagged with graffiti. It reminds me of *RoboCop.* We pull up in the predawn dark. The place where dreams are made.

I sign a confidentiality agreement. I am an initiate, an outsider. Worse, I am that most suspect of villains: the writer. I will become privy to secrets that won't be known until season two begins airing in a few months. Because I won't betray my friend—who got me on the set—and because these people are masters of the double cross, the legal ensnarement, the ruination of men, I must be circumspect.

I meet the director in an interrogation room, bare save a table and two metal chairs. A single light burns above. A sign warns us we're under video surveillance; badges must be worn at all times. The walls press in, thick, gray cinder blocks. Thump them with your palm. Hollow. Painted. On the other side of the door, the crew grumbles about yesterday's Cowboys game. Six turnovers, four interceptions. Should they bench the quarterback? They're eating chilaquiles with

Mexican peppers, hot off the breakfast truck. What did you do last weekend? I cut down cottonwoods. The director and the writer discuss a line in Urdu. Nearby, blue light streams through hospital windows. The swimming pool murmurs at Southfork, where painted pastures stretch for miles. A yellow-and-white-striped awning ripples in the breeze. An elevator door opens onto the twenty-eighth floor. Step into the offices of Ewing Energies, where someone is playing a radio. A note on a secretary's desk reads "There's no business like it!" The windows command a sweeping view of downtown Dallas, a massive neon green screen. The sun is not up yet but today is bright and sunny and unseasonably cold. A woman is washing a window to nowhere. We are hermetically sealed. Cliff Barnes—J. R.'s nemesis—steps out of a red golf cart smoking a cigarette. A hush falls. Quiet. Private rehearsal. A hallway in the hospital turns seamlessly into a police station with wanted posters on the wall. I recognize some of the faces on the posters working around me, manning the cameras, the props, the lights. Next to the posters stands a biohazard box; a public-health sign declares SOME OF THE BEST MOTHERS ARE FATHERS. Hospital bleeds into police station which bleeds into jail. Twelve of us crowd into the interrogation room. A man snarls at me, "You set me up!" Then the walls are torn apart to make room for lights.

A crowd gathers in the hallway of the police station. The director and writer debate the feasibility of staging a traffic jam today. A guy leans against a wall that suddenly gives way. "Who built this set?" he asks. The director says to no one, "Sometimes you forget it's not real." Back in the interrogation room, a man enters in an orange jumpsuit. My friend, the writer, has killed him, and he is upset. Someone suggests a spin-off. The grip next to me says, "I'm on board. I've been making up stories about this show since I was twelve."

Announcement: There will be policemen on set today. All guns, handcuffs, and pepper spray canisters are fake.

The scene is repeated over and over: a father loses a son; the son loses a father—and an empire. They speak the language of guilt, betrayal, duty. With headphones on, we sit in the dark and watch. Reset the scene. A woman laughs, a hammer bangs, something heavy clatters to the floor. Then the bell rings, the great red light is lit, we're rolling, the director calls, "Action!"—and everything and everyone is suddenly still, all eyes on the monitors. One camera shoots from above; another glides on rails up and down the room at a glacial pace, imperceptibly heightening the emotion, making one trip there and back as the actors cycle through rage, remorse, denial, and defeat. We restart when the boom mike drops into frame. The master is finished; now for the close-ups. The actors are ringed in light. They're beautiful. Then they're gone, replaced by doubles of roughly the same height wearing roughly the same clothes—as if seen through a fogged mirror—who pantomime the scene while the crew rehangs the lights. Shooting the same scene from all possible angles is called *coverage.* It erases mistakes. It's what we all wish for. Now tight on the feet and sweep up. Once on the door as it opens. Once on the clasped hands. Over and over for more than two hours, we watch two men in a box choke themselves up again and again. My headset puts me in the room where I can hear the smallest whisper. A dozen cops walk by, unnoticed.

In the police department, Christopher Ewing fights for custody of his babies. I know the man next to him, his lawyer, as a secret-service agent on another TV show. There is some walk and talk, some Steadicam. Nearby, seated at breakfast tables outside the set, a corps of cops is receiving instruction. "Find a prop and carry it, if possible." Away from the action, it is cold and dark. There is no time and no weather. In one corner, a giant tree reaches for the ribbed ceiling, neglected, forgotten. Outside, in the parking lot, nine cars are being arrayed around a $350,000 Rolls-Royce to give the illusion of a mid-day traffic jam. Inside, the prop master approaches the writer with a question about the color scheme of a fake search engine that churns out false results. The writer likes what he sees. The script supervisor marks her pages, noting the continuity.

In Ewing Energies, Christopher and John Ross—the sons of Bobby and J. R.—square off like their fathers. They walk; they lie. The camera picks them up, pivots around them, tracking their do-si-do of deception. "They're the center of their own storm," says the director. Elsewhere, Sandy is massing strength and about to smash into New Jersey. The winds have reached ninety miles per hour. Obama will speak soon. We're in a bunker.

I drink a soda in the Ewing Energies kitchen. I have to keep myself from throwing the empty away in the prop trash can. The fridge holds huge lightbulbs. The office is a $400,000 fake. A crew member hums the *Dallas* theme. There is a problem with the choreography: a line is needed to buy Christopher enough time to walk over to John Ross, who is talking to an oil hand. Just a throwaway—something small. The writer looks at me. I have five minutes to think. I give him the words. The director signs off. And suddenly I've become history. I feel a giddy thrill. I've written my small bit into *Dallas* lore, which—depending on which take they use, and if that take makes the final cut—will amount to four or five words no one will notice.

Another scene at Ewing Energies: four cops exit the elevator. Three of the four are real Dallas PD. Are *their* guns real?

After a thirteen-hour day, I'm at my hotel, eating alone at the Rattlesnake Bar. Next to me, two men talk golf. One of them has just moved to town. The local tells the new guy a long story that ends with, "He asked me to cash his check, and it was a royalty check from Texaco!" The men laugh.

I wake up at 5:00 a.m. to watch Judith Light administer a sponge bath. I grew up with her as the lovable mom on *Who's the Boss?*, which ran from 1984 to 1992. Now she's a cruel, manipulative matriarch. But her son is wounded, so Judith Light cries and cries over a man pretending to be in a coma. Then the man rises and leaves behind the hospital machines working so hard to keep him alive.

Back at Southfork, I witness Christopher drink enough bourbon to kill a horse, take after take. But it has no effect. Each time, it's as if he's having the first bracing drink of the day. This is a land of no hangovers, physical or spiritual, a place designed for the coupling and recoupling of impossibly attractive people. A wonderland. The fountain of youth. But there is a price. Someone tells me they think each episode runs north of three million dollars.

We—the director, the writer, and others—sit just off the set in what's called "video village." The actors are in another room; currently, we are blocked from seeing the real thing by walls, scrims, screens, and a closet full of shotguns. But we see it better than they do—from more than one angle on twin high-definition monitors. If it doesn't happen on the monitor, it doesn't really happen. I turn off my headphones. I faintly hear the voices from another room. Sound bleeds from the director's headset. Am I crazy, or is there a lag—are we, on the headphones, just slightly in the past?

Historical events used to be indexed by markers and monuments—physical objects that pointed to, but did not literally depict, the events that happened. Thus tourists go to Gettysburg and marvel at granite slabs that stand in for the regiments that fell on the field. End of story. Not so for more modern history, where imaging technology has become so ubiquitous and persuasive that *nothing* ever ends. We experience a reality that is endlessly replayable and remixable—on film, on TV, on cell phones, online. Here is how history unfolds today: we watch the event in real time (often on a screen in front of us), and then the footage is shuffled off to a digital cloud, where it can be accessed at any time from anywhere. As I type this, on a screen somewhere, the muddy waves of Superstorm Sandy slam into Atlantic

City, Predator drones rain bombs down on Middle Eastern deserts, and the Twin Towers again and again crumble to dust. These videos are not mute artifacts of the past; they *are* the past, happening right before our eyes.

Nothing ever ends. It is only a click away. And because events never end—and because the images themselves offer no clear and stable interpretation, but instead can be endlessly dissected and debated—the past, as the saying goes, is never really past. Because we can witness the past, we think we can know it—but we can't, not fully. The footage remains grainy, no matter how close we get. And that frustrated sense of knowing/not-knowing leaves us stuck, on the verge of an illumination that never arrives. We can't put anything behind us.

Where was JFK assassinated? For most of the country, Kennedy didn't die in Dallas, but on television. And that's what Oliver Stone got right in his oft-maligned 1991 movie *JFK*—since the release of the Zapruder film, we, as a country, have been stuck in a dark room, watching history being endlessly replayed, back and to the left, back and to the left . . .

And so when people come to Dallas for the ultimate close-up, they come not to the DFW metroplex but to Dallas-on-the-TV. They're here not to understand history, but to experience it. To stage it. To shoot it. To bend the physical reality to the collective fantasy—and then insert themselves in it. Through equal parts erasure and reinvention, Dallas has made itself a blank screen, a place for others to project their opinions and for the city to sell its story.

Meanings pile up; they fold, they fault. Pressure builds. This dissonance comes at a psychic cost. Kennedy's body is buried in Arlington National Cemetery, but Dallas bears the wound. Growing up, for me Dealey Plaza was a blank spot—a confusing interchange of two- and one-way streets on the edge of downtown, a dark patch of grass you might mistakenly drive past on the way to a basketball game at Reunion Arena (now gone), a place not to be lingered in, a footnote left unread. I never visited Dealey Plaza on a school field trip. I never went there with family. In the years after the shooting, the sixth floor of the Book Depository remained empty—with county business eventually taking place in the five floors below. A public exhibit didn't open until 1989.

It takes me a map to find it.

We stand on the ground floor of the Dallas County Administration Building, formerly known as the Texas School Book Depository, waiting to pay sixteen dollars. The woman selling tickets asks, "What zip code are you from?" I tell her, and she says, "Thanks, Sugar."

Six floors up, a parent addresses a group of ten or so children: "Go learn something. Have fun!" The kids hit the green button on their audio tours and disappear into the Sixth Floor Museum, where there is no food or photography allowed, and cell phones are to be kept silent.

I have traded one set of headphones for another—the audio feed for the audio tour, one kind of simulated reality for another. This tour is historic, contextual, stitched together with eyewitness accounts. A voice gives me instructions on how to move through the space—where to go, how to look. There are Germans, schoolchildren, and Japanese tourists with clipboards. We form an uneasy procession, alone together.

What we learn: Only forty-two years old when he hit the campaign trail, JFK was a young leader for a new generation, the television president, as Nixon learned so disastrously at the debates, unprecedentedly photogenic and media savvy. JFK used news photos, press conferences, TV appearances, and glossy interviews to produce his own reality: Camelot. But his presidency wasn't built on thin air; it intersected with Big Historical Events: the Civil Rights Movement, the Cold War, Vietnam, the Space Race, the Peace Corps, the Cuban Missile Crisis, the Nuclear Test Ban Treaty, among others.

JFK, an Incomplete Timeline

November 4, 1960: Lyndon and Lady Bird Johnson—campaigning in Dallas on the eve of the election—are harassed by a mob of three hundred Republican women dressed in red, white, and blue, who tear off Lady Bird's gloves and, as the Johnsons attempt to enter the Adolphus Hotel, block their way and spit on them. Johnson slows down and—for the benefit of the cameras—suffers the abuse for an excruciating thirty minutes. Civic leaders express shame, and the state swings for Kennedy and Johnson. But the idea of Dallas as a conservative vipers' nest is born.

October 24, 1963: The ambassador to the United Nations, Adlai Stevenson, is whacked on the head with a placard wielded by a forty-seven-year-old woman wearing pearls. City leaders issue a telegram of apology on behalf of the city; they send a copy to President Kennedy.

November 21, 1963: About five thousand anti-Kennedy handbills are distributed around Dallas. "Wanted for Treason." That evening, the *Dallas Times Herald* publishes the route of Kennedy's motorcade on its front page.

Friday morning, November 22: A full-page ad appears in the *Dallas Morning News* lambasting Kennedy and his policies. It is paid for by an anonymous group of right-wing businessmen calling itself "The American Fact-Finding Committee." Many Dallasites find the ad in poor taste, even rude. It appears on a day that will spawn countless committees searching for the facts.

November 22, 11:37 a.m.: Kennedy lands at Love Field. He shakes hands across a wire fence with the rabid and joyful crowd there to greet him.

A video shows the moments leading up to the assassination. The footage is familiar, the frames iconic. The big welcome for the president. His motorcade creeping toward Dealey Plaza. Jackie in a pink wool suit with navy collar and matching pillbox hat. Five kids cluster before the monitor, holding their breath as the limo glides across the plaza. Just before the first shot, the frame freezes and bleeds into white. A boy whispers nervously to a friend, *"Did you see it?"*

Jackie's rose bouclé suit was a copy, or re-creation, made by an American dressmaker of a Chanel original. Jackie wore it that day at her husband's request. The dressmaker made it at her request. Jackie needed to wear something made in America.

The exhibit traces the motorcade's route downtown. Some ten thousand supporters line the streets. The wind catches Jackie's hat, and she reaches to steady it. The limo rolls down Main Street, then up Houston, passing the jail to which Oswald would have been moved had he himself not been assassinated. We wheel left through the torturous hairpin turn onto Elm Street, passing below the windows of the Book Depository. Nellie Connally, the governor's wife, says, "Mr. President, you can't say Dallas doesn't love you!"

First, a Polaroid blown up to an enormous size: the image is grainy, impressionistic, a blurry pointillist blot. A voice in our ear guides us to stills from the Zapruder film pasted on the wall. Seven slices of life that occurred between 12:30 p.m. and 12:31 p.m., as recorded onto

8mm Kodachrome II color film moving roughly at 18.3 frames per second through a Bell & Howell Model 414PD Zoomatic Director Series camera with a Varamat zoom lens trained on full close-up on a subject in a moving car some sixty-five feet away.

The first frame at 12:30.

Then 1.26 seconds later.

Then 2.23 seconds later.

Then .7 seconds later.

Then .9 seconds later.

Then 1 second later.

Then 2.1 seconds later.

One of the stills is the achingly familiar Zapruder frame #238, for many casual viewers the first sign something's wrong: Kennedy's hands go up to his throat, elbows out like wings. After, his chin drops, he leans over, then comes the head shot, which sends him—as everyone knows—"back and to the left," and he topples into the seat and Jackie climbs over the rear of the limousine, scrambling for a piece of her husband's brain.

After the third shot rang out, Jackie shouted, "They've shot him." Governor Connally, wounded by the second shot, said, "My God, they are going to kill us all." Both of them already supposing a conspiracy, a "they."

The day after the assassination, Abraham Zapruder sold his film to *Life* magazine. The magazine damaged the film, leading to the loss of six frames—but even Oliver Stone doesn't take this to be a sign of a cover-up. On November 29, *Life* published thirty-one black-and-white frames of the film, plus another nine in color two weeks later. Americans pored over the photographs.

A less famous film, taken by Marie Muchmore and showing the shooting from another angle, was bought by United Press International and aired in its entirety on WNEW-TV in New York on November 26, four days after the assassination.

As the limousine races to Parkland Hospital, it speeds past the Trade Mart, where people awaiting a luncheon with the president now bow their heads and pray for him. At the head table, a place setting with a delicate coniferous motif sits empty.

Next exhibit: a map of the route the victims took through the ER to reach Trauma Rooms #1 and #2 in Parkland. A dotted line winds through the halls, then splits at the rooms. Red dots represent JFK (room #1), Connally (room #2, across the hall), and LBJ (waiting in Minor Medicine).

While her husband is in the hospital, Nellie Connally delivers a statement on a telecast: "It had been a wonderful tour and when we arrived in Dallas, and were in the motorcade, the people could not have been friendlier, the crowd more wonderful or more generous in their reaction to the president. The city of Dallas does not deserve to be blamed for this ghastly crime."

Years ago, late one night at a dinner party in New York City, I found myself seated across from a tipsy older socialite, who—it was revealed—had grown up down the street from my grandparents in Dallas. She remembered little things: what the inside of their house looked like, that my grandfather—a vascular surgeon—patched up the kids and pets of the neighborhood. She then raised her glass of wine and dropped a bombshell: "Oh, he was such a good surgeon! You know, the whole block knew he operated on Kennedy, but of course no one talked about that!" I was flabbergasted; my wife couldn't believe I didn't know the story. I called my mom the next morning and she assured me that—despite the woman indeed being an old neighbor—my grandfather hadn't operated on the dying president. But the woman had one thing right: no one talked about it.

Behold the sniper's nest: a glassed-in exhibit of the fatal sixth-floor windows. The walls are brick, the floor wood. Boxes of books block the southeast corner window. The adjacent window on the south wall is half open; boxes stacked before it serve as a seat. The voice in my ear explains this is a "reconstructed display" of how detectives found the sniper's nest minutes after the shooting, and, indeed, what I see matches the evidence photo on the panel in front of me. Detectives found three cartridges below the window and Oswald's fingerprints everywhere.

While I'm peering through the glass at the display, a woman next to me tells her kids: "None of this has been touched in, what, forty years?" I don't correct her math or her assumption, but let the silence linger, as they confuse the re-creation with the real.

We line up along the south windows and imagine the path of the motorcade. A man in a baseball jacket stabs his finger at the glass: "I don't understand why he didn't shoot him there—it's a much easier shot!" "That gun was known for jamming," a woman offers. "Perhaps it jammed." Their son asks, "But who shot him though?" and the question goes ignored.

Below us, somewhat inconceivably, a film crew is setting up. They carry large back scrims. A red car of unknown vintage pulls up. We see an actor in a dark suit. Next to him, a flash of pink. An excited woman asks, "Are they re-creating it?"

Meanwhile, white *X*s dot the middle lane of Elm Street, supposedly marking where Kennedy was when the bullets hit. No doubt I've driven over them countless times in my life, unnoticed. A couple takes a picture on the green of Dealey Plaza, posing in front of the Book Depository. Behind them looms Reunion Tower. The traffic breaks, and a woman in heels ludicrously teeters into the middle of the street to pose on an *X*, putting herself firmly in the crosshairs of history as an accomplice shoots a picture.

Stuck at the windows, I've left the audio guide playing, and now we're on to Oswald's apprehension and assassination in the basement of

Dallas police headquarters. Strip-club owner Jack Ruby would plead not guilty by reason of "psychomotor epilepsy," claiming he was in a sort of unconscious dream state when he shot Oswald. The jury sentenced him to death by electric chair, but he appealed, and it was cancer that killed Ruby in 1967.

Back at the studio, the Dallas police seem a very companionable lot, not threatening at all as they stand around and eat their yogurt.

The Secret Service reenacted the shooting in late November and early December of 1963. The FBI reenacted the assassination on May 24, 1964. I stand before a model of Dealey Plaza built by the Bureau; a quarter of an inch equals one foot and white strings trace the trajectories of the three shots. The evidence is continually sliced and diced. A glass case holds twelve cameras, film and still, that captured the event, with a blown-up image from each. I study a blurb about the 1976–1978 acoustic investigation that proved with 95 percent certainty the existence of a second gunman; I read about the 1980–1982 counter-investigation that contradicted the previous investigation's certainty. Doubts and mysteries pile up: "the back wound" that doesn't match the autopsy report, "the pristine bullet" found on the gurney at Parkland, the curious dent in the limo and chip on the curb. Culprits abound: the Soviets, the mob, Texas businessmen, the FBI, the CIA, Fidel Castro, anti-Castro groups, and so on. In 1981, Oswald's body was exhumed in an effort to put to rest theories that a double had been buried in his grave. In 2003, a survey found seven out of ten Americans still believed in a conspiracy. I compare a map of the microphone placement used for reenactments to one tracing the normal "echo patterns" in urban environments. I mouth the words "echo patterns."

I grew up in the echo pattern. I didn't live through the event. There are people whose firsthand experience leads them to regard this story as a simple—albeit horrible—one about fear, hate, derangement, evil—

take your pick. For them, my story, built on absence, might seem pointlessly complex, obdurate, even obscene. But that's the problem with echoes: they don't tell you where to look. Despite the museum's levelheaded, authoritative tone, we stitch our own stories. We linger in front of the panels we want to believe: theory, counter-theory, fact, supposition. To wander the museum is to wander through a dark day, to wander through doubt and uncertainty, to wander through the mystery of why men murder.

At the end of the exhibit, we are invited to record our thoughts in the "Book of Memories." I leaf through the pages. There are lots of sad faces and doodles of the initials JFK. (Kennedy himself was an incessant doodler.) Someone writes, "I wish that he was still with us." In cursive: "John F. Kennedy is the best." Keep flipping: "I'm Jess and I think Lee Harvey Oswald worked for Jack Ruby and planned all this." "Thank you for your service, JFK." "RIP JFK." A visitor from Nepal: "Greatest man ever." A line of Chinese characters. "A wonderful walk through history to honor such a man." "What a meaningful tribute."

Unable to put my finger on the meaning, I get into the empty elevator. A man enters at the last minute. As we descend, he breaks the silence: "That's a lot of history in one place."

Exiting the museum, I step onto the plaza and push past two men hawking conspiracy newspapers. They act suspiciously, either to drum up business or dodge the authorities. But the cops are busy holding back traffic and pedestrians as the film crew gets in place. In front of me, a man in a dark suit and skinny tie crosses the street with a young raven-haired woman in a pink wool suit. There are dark stains on her lap.

We, as a crowd, stand dazed, hearing echoes, as the camera gets its shot. People start taking pictures of the woman, as if this too were something to be documented and dissected. She seems sheepish, even guilty. She puts on a coat and hides behind the man; then she disappears.

J. R. will be here soon, someone tells me back at the studio. And so I wait. In the late afternoon, there is a strange moment when the light streaming through the living room windows of the ranch actually matches what it would be if we were outside. The set is momentarily in sync with the world. Artifice equals reality. My internal clock chimes true. But time slips on—outside, the sun sinks—while we remain stuck in the perpetual late afternoon, waiting for J. R. to arrive.

Suddenly, I turn around and there—lost in a crowded room—is J. R., tying his own tie. He fidgets with the knot; he cannot get it right. No one helps him until a prop person appears with his hat.

J. R. is not himself. He sits between Bobby and Sue Ellen, joking like old friends. They kiss, they hug, they give each other grief. They're shooting a promo for the show. They ad lib, mug for each other, steal one another's lines. While Sue Ellen fixes her hair, J. R. pretends to fall asleep. Bobby gives him an elbow and J. R. comes up sputtering. It's a vaudeville routine, not a blood feud going back decades. J. R. is the life of the party, but, again, he is not acting like himself. He is too happy, too in love with his brother and his ex-wife. Between takes, J. R. gives me a fist bump—a fist bump!—and offers some career advice. "Write, write, write," he tells me. He is warm and caring. Maybe I have been entirely wrong about him.

The jacket, hat, and shoes Jack Ruby wore on the day he shot Oswald are in an auto museum in Roscoe, Illinois. The toe tag from Oswald's corpse has been sold a number of times to private collectors. Kennedy relics are harder to come by. The entire contents of Trauma Room #1 at Parkland are interred in an underground government storage facility in Lenexa, Kansas, and are not accessible to the public. Kennedy's limousine was dismantled to its frame, then rebuilt and used by Johnson, Nixon, Ford, and Carter before winding up in the Henry Ford Museum in Dearborn, Michigan. Jackie's bloodstained pink suit is stored at the National Archives, with the consideration that it never be publicly displayed, an agreement that is up for renegotiation with the family in 2103. Currently, the suit rests in a box in an undisclosed room whose air—kept between sixty-five and sixty-eight degrees and at 40 percent humidity—is recycled every ten minutes. The pillbox hat is missing, presumably lost, possibly sold—though no one is sure. Three years after the assassination, the bronze, satin-lined casket used to transport Kennedy's corpse to Washington was bored with holes, filled with eighty-pound bags of sand, put into a pine box that was also bored with holes, and dropped by the Air Force into a section of the Atlantic Ocean off the Maryland–Delaware coast used as a weapons graveyard, where—because of unexploded ordnance—it cannot be recovered, an empty coffin within a coffin, buried nine thousand feet beneath the waves.

At 4:20 p.m. on November 23, 2012, fewer than four weeks after I saw him—and forty-nine years and a day after JFK's assassination—Larry Hagman died in a different Dallas hospital with his fictitious brother, Bobby, and ex-wife, Sue Ellen, at his side. The world went into mourning: celebrities took to Twitter; the BBC called Dallas night and day for reactions; fans flooded Southfork. Hagman had finished shooting five of the scheduled fifteen episodes of the new season of the show; the producers assured fans J. R. would be given the send-off he deserved. As for Hagman, he had not one but two private funerals, one in Santa Monica and one at Southfork, full of real and fictive family. A day after the Texas funeral, fifteen hundred fans from across the world flooded the ranch for an open memorial, where they took tours, left mementos, and signed a book of memories. In 1988, Hagman announced what he thought should go on J. R.'s tombstone when his time came: "Here lies upright citizen J. R. Ewing. This is the only deal he ever lost." Hagman has no burial site; instead, his son is said to be scattering his ashes around the world.

Dallas is a rich man with a death wish in his eye . . .
A rich man who tends to believe in his own lies.
—JIMMIE DALE GILMORE, FROM THE SONG "DALLAS"

The last shot of the day. Night: J. R. sits in a chair in his room, alone, scheming. A single lamp burns. A drink on the table. The phone rings, a nefarious call—lines of influence are crumbling, webs tangling. J. R. speaks three lines into the phone, going from vindication to rage. There is still a wall between us, but I've got my headphones on. The camera lingers on his face as he stares into the middle distance. Behind his eyes, deep oceans are churning. A fire burns. The script suggests a fuse has been lit. The man next to me shudders. We all shudder. Because, suddenly, there is J. R. He's there on TV, where we always can find him, where we should have known to look.

Four months later, this scene—which will have already aired once on TV—will be digitally altered (the walls turned red, the audio looped, the lips matched, the whole thing sliced and spliced) so that it becomes something new: J. R.'s death scene, his final appearance on the show. And who shot J. R.? Long odds at last come in: it turns out his death *was* a suicide; he was terminally ill, but he had a henchman shoot him so he could frame his enemies from beyond the grave. But this is not, of course, his final appearance, any more than it is an *actual* appearance. Rather it is a haunting, the echo of a ghost.

A block away from the Book Depository, in a plaza next to the County Records Building, stands John Neely Bryan's cabin—or, rather, its replica. A plaque at the foot of the cabin gives an inaccurate account of how Dallas got its name. The cabin's door and windows are padlocked. You can't see inside.

The simple wooden door faces the John F. Kennedy Memorial in another plaza across the street. The memorial, designed by architect Philip Johnson and dedicated in 1970, is a thirty-foot-high open cube of white precast concrete; perched on eight legs, it floats twenty-nine inches above the ground. The memorial is a cenotaph, or "empty tomb." Two narrow slits face north and south; the tomb remains open to the sky. An inscription on the ground reads: "The death bullets were fired two hundred yards west of this site."

The cenotaph is meant to evoke the freedom of Kennedy's spirit.
It is empty, save for a black slab upon which a body might rest.
But there is no body.
Nobody.
Nobody inside it but us.

2012–2013

Lost and Found

In Gettysburg, Pennsylvania, a few blocks up from the town square, stands a three-story redbrick house with green shutters and white trim. A porch wraps around the front. A small white cupola perches on top like a dollop of frosting, from which nervous members of my father's family used to watch for German bombers during World War Two. Today, the house is owned by my father and his two brothers. An uncle of mine still lives there, and my parents visit for three or four months out of the year.

My great-great-grandfather, Edward McPherson, built the house in 1870, seven years after the battle that made the small orchard town famous. I am named after him, as is my father. What do we call the house? An "ancestral home" sounds too grand. A "homestead" sounds like it's out on the prairie. No, in our family, we refer to it as "Gettysburg," somewhat presumptuously, as if a house can contain a whole town.

Thanks to a series of generations that believed in staying put, generations of dutiful men, men of obligations, who might briefly engage themselves in the business of larger towns, say Washington,

D.C. (only an hour and a half away), before returning to occupy the house of their fathers—thanks to these men and their dutiful wives (who were expected to play their part and follow along), plus a lonesome collection of distant maiden aunts, the house is stuffed with artifacts, none of them organized, everything crammed together under its peaked gray roof.

Things I have found in the house:

- Bullets picked up from fields and farms shortly after the battle.
- Artillery shells and cannonballs that, I trust, have been emptied of explosives.
- A military sword of unknown origin that might or might not date to the Civil War.
- Maps of the battle and the surrounding communities, plus a layout of the family pews in the Upper Marsh Creek Presbyterian Church, dating from the late 1700s.
- The framed commission papers of Colonel Robert McPherson, who served in George Washington's Continental Army.
- A young girl's diary from 1864, offering candid views of Philadelphia and Chambersburg during the Civil War.
- A copy of *Don Quixote* from 1865, with illustrations by Doré.
- A History of Lancaster County, Pennsylvania, from 1883.
- A massive, leather-bound copy of the mystical devotional *Vier Bücher vom wahren Christenthum (Four Books on True Christianity),* by the Lutheran theologian Johann Arndt, printed in a heavy Gothic font and dated 1733 on the title page.
- A chest containing some quilts that, according to an accompanying note, were singed when the Rebels burned Chambersburg.
- Envelopes stuffed with green and red one- and two-cent stamps from the 1920s.

- Reels of home movies, including footage of my great-grandfather watering the garden (wearing a suit) in 1935, of Ike's inaugural parade in 1953, of the 1960 Summer Olympics in Rome.
- A collection of Agatha Christie paperbacks.
- A copy of the self-help book *I'm OK, You're OK.*

A shorter—and more fraught—list of things I have seen go missing over the years:

- A number of early editions of L. Frank Baum's *Wizard of Oz* series, containing illustrations I used to stare at for hours. (Culprit: a generous and shortsighted grandmother, clearing out space. Recipient: the Salvation Army.)
- A collection of P. G. Wodehouse hardcovers given to my grandfather as a boy and which, as an adult, I was working my way through. (Culprit: an uncle, also clearing out space. Recipient: the local library, from which the books promptly disappeared.)
- And, perhaps the *pièce de résistance,* Edward McPherson's Civil War sword. (Culprit: grandmother again, for reasons no one can fathom.)
- It is selfish to hang on to what you cannot use. I still harbor a grudge.

Three things I have taken from the house without asking:

- An article from a local paper announcing the graduation of my grandmother from Yale Law School in 1935. It reminds me of the panoramic photo I found in a closet on the third floor—now on display in the TV room—of my grandmother with her law-school class. Rows and rows of sober men in dark suits stand with arms folded in front of an ivy-draped building. There, in the second row on the right, the only woman in the frame, wearing a brilliant white frock, is my grinning grandmother. (Despite her daring in that time and place, she didn't practice law for long. Four years later, in a Yale University chapel, she married a lawyer and gave up bustling New Haven for Gettysburg.)

- Two miniature books, bound and printed in London by the Collins' Clear-Type Press (undated, but old and crumbling): *Gems from Shakespeare (Comedy)* and *Gems from Shakespeare (Tragedy). Comedy* is in good condition, but the binding is broken on *Tragedy.*

When he was in his twenties, my father met and married my mother, a good Texas girl. Together, they lived in Washington, D.C., for a brief stint, and then, instead of returning to Gettysburg, dropped below the Mason–Dixon Line, citing expanding business opportunities and a far better climate in Dallas. My grandparents never really understood the decision. Pictures show them—the father and mother of the groom—looking happy but slightly dazed at the wedding. Perhaps they are surprised to be sitting outdoors by a pool in early May. They would visit us at Christmas, and we would return to Gettysburg for a few weeks every summer. I was seven the night my father was summoned to my grandfather's deathbed. I remember the phone call—from my grandmother, telling him the time was near—and the light coming from under my dad's closed office door, which I could see from my room, burning on through the night.

When I was in college, my family began spending Thanksgivings in Gettysburg—it was an easier trip for my sister and me, now both on the East Coast, and we continued the tradition even after my grandmother died. She'd been gone for two years when I first brought my future wife to Gettysburg. Five years after that, we were married in a church half a block from the old house.

I try to get back at least once a year but return more often in my writing. I shoehorned a chapter about Gettysburg into a nonfiction book I wrote about bridge, the tenuous link being my grandmother's love for the game. I wrote a short story about a man who paints cycloramas—massive, 360-degree canvases designed to literally envelop a viewer within a historical scene. They were wildly popular at the end of the nineteenth century. The virtual reality of their time, cycloramas represented the pinnacle of science, entertainment, and art. Everyone knows the one in Gettysburg depicting Pickett's Charge. I have walked behind the heavy linen canvas and marveled at the hyperbolic curve that gives the painting its curious perspective.

Recently, I wrote a short story about a reunion of Civil War veterans. For research, I read a *New York Times* article from 1891 regarding a kerfuffle over the placement of a battlefield monument. The men of one regiment wanted their memorial placed at the front line of Pickett's Charge, where they claimed they had fought. However, veterans of other units said the regiment had been slow to take its position—that the soldiers had, in fact, hesitated when given the order to engage. Two sides emerged, telling different stories of where, exactly, the men had stood twenty-eight years earlier. According to the article, a resulting legal battle over the monument made it all the way to the State Supreme Court. I experienced a strange jolt when I came upon this sentence: "Mr. Edward McPherson of Gettysburg, in accepting the monument for the Battlefield Association, said he hoped the time would come when historic truth would triumph." It was like suddenly catching a moving reflection in a mirror. I didn't know my great-great-grandfather had been involved in policing historical markers. He was a confident fellow, and I am jealous of his optimism, however guarded.

Edward, who built the house in Gettysburg, was a newspaperman, attorney, and political junkie. An acolyte of Thaddeus Stevens, best remembered as the architect of Reconstruction, Edward was himself elected to the House in the thirty-sixth and thirty-seventh Congresses

(1859–1863). After losing a bid for reelection, he occupied other government posts, edited several newspapers and journals, and wrote thick books with long titles like *A Political History of the United States of America During the Great Rebellion.*

He liked to write speeches, copies of which abound in the house. Some of these he presented to the men of the local YMCA; others he delivered in Washington. He was a good Lincoln Republican, and on February 14, 1862, delivered an address to Congress called "The Rebellion: Our Relations and Duties." The next year, he spoke to a local college on the topic "Know Thyself." He had just lost an election by ninety-two votes. Perhaps it was a time of soul-searching.

When I go to Gettysburg, I am torn between visiting with my family and digging through the house for old forgotten things. A few years ago, I found a framed ink drawing of a geometrical starburst pattern that surrounded a string of curious characters made up of right angles and dots. A website hosted by the CIA helped me crack the code, an old Masonic cipher. According to the text, the drawing was made in memory of a man named Godfrey Lenhart (1754–1819). He was Edward's grandfather, my great-great-great-great-grandfather, and a clockmaker, silversmith, and, apparently, a Mason.

One summer, I found a small wooden box stashed away in an overlooked drawer. The box contained a thin piece of thread. A handwritten note explained that it had been clipped from Lincoln's funeral catafalque, the cloth-covered bier on which his coffin had rested. It was short, a black little worm, something you might brush off your sleeve. A few minutes on the internet revealed that they were given as mementos to members of Congress. At the time of the assassination, Edward was the Clerk of the House of Representatives. The thread no longer lives in the house, but in a safe-deposit box at the bank.

A shoe box on the third floor holds a packet of papers tied together with string: letters my grandfather's brother sent home during the summer of 1917, shortly before he died in a waterfront accident at camp. My grandfather, who was ten, was at camp with him when it happened. His letters are there, too, and stop—like his brother's—the day before the accident. My grandfather never said much about it. I had always heard his brother drowned while swimming in a quarry, but a yellowed clipping from the local newspaper says the boy died when a riverbank cave he was exploring suddenly collapsed.

At the age of six, my father was sent to camp for a startling eight weeks. His father would send him letters typed by his secretary. My father says he loved every minute, and returns periodically for reunions of campers and counselors. I went to camp, too, but for only four weeks and at a much later age. The letters I received were handwritten. I also loved camp. From no letters to typed letters to handwritten notes—perhaps that is progress. My father's family is not known for open communication. Between generations, things don't get passed down; they accumulate.

I don't like to throw anything away. My parents are unhappy about this. From Texas, they threatened to ship me boxes of small rocks, broken karate boards, and rusty machine parts. One fall, they put the boxes in the back of their car and drove six hundred miles to deliver them to me. I have almost every personal letter I've ever received. For years, what used to be my closet in Dallas held old T-shirts and a dinky Irish tin whistle I can still sort of play. Now that stuff sits in a box, too.

I worry that I'm indulging myself in that cliché about undemonstrative men, those bankers of the heart. My father was a three-sport athlete at boarding school and led the basketball and baseball teams in college. He told me that his father only made it to a handful of his games—certainly none of the big ones in college.

My father works in finance; he collects books on business and often overdresses for the occasion. (His father appeared at the dinner table every night in a suit.) But he saw most of my high school games and even kept track of my collegiate soccer career, which took place on the other side of the country. Growing up, he didn't coach me from afar or yell at the ref. I had plenty of space. Sometimes he would sit alone, high in the stands, away from the other parents—a solitary figure in a long, black dress coat. I wouldn't always know he was there. I remember warming up for an away game during my freshman year of college and having a teammate from Greece point to the top of a hill and say, "Hey, isn't that your dad up there?" I told Angelo to get his geography straight—Texas was seventeen hundred miles away from where we were in western Massachusetts. But he was right. After the game, my father approached me in the parking lot and explained, "I was in Boston on business." A few minutes of small talk, and he left. Boston was more than two hours away.

In Gettysburg, the past is both intimate and distant. The house is so full—and so fragile. Things keep getting lost. How to unpack it in a lifetime? Who can or even wants to sort everything out?

In a drawer of a file cabinet, I find a number of land patents written in cursive on some sort of heavy cured parchment. They have wax seals attached by faded blue ribbons. I open a deed announcing that the proprietors and governors of the province of Pennsylvania allot five hundred acres in Lancaster County to one of my distant ancestors, "as recorded on the 19th day of March 1743 and witnessed by the hand of . . ." The name has faded out.

Three brothers now split the cost of keeping up the house. Recently, in one year alone, it needed a boiler, some plumbing, and a new roof. Among the brothers, there are four children—not one of us has the flexibility or the financial means to take over. We are scattered across the country. There could come a day when no one in my family lives in Gettysburg.

The house, with its almost unbearable, unknowable inventory, has suffered some sizable amputations over the years. A good chunk of Edward McPherson's papers—some eighteen thousand items, filling one hundred containers and one oversize bin—was donated by my grandfather and great-grandfather to the Library of Congress and now occupies 40.4 linear feet of shelving in Washington, D.C. Edward was a genealogist, historian, archivist, and packrat himself, who collected material from *his* ancestors at least four generations back: ledgers and ciphering books, powers of attorney, estate accounts, a "horse stud fee book" from 1771 and 1772, a list of dogs taxed in Gettysburg from 1806 to 1816. And then there's the accumulation of Edward's own life: maps, calling cards, notes for speeches and books, obituaries, invitations, correspondence (from James A. Garfield, from Stevens, from Lincoln), his college report cards, the minute book of the Shakespeare Club in 1847, and his two-volume scrapbook on the Battle of Gettysburg. On cold, damp days, when the bookshelves creak, I imagine the house is registering their absence like a phantom limb.

In his essay "The Catastrophe of Success," Tennessee Williams writes, "The monosyllable of the clock is Loss, Loss, Loss, unless you devote your heart to its opposition." The past is both relentless and inarticulate. How can you argue with a monosyllabic opponent? I am starting to sound maudlin, so it is important to remember that Williams also saw the lighter side of things. He struggled as a screenwriter in Hollywood, and when MGM asked him to rewrite a B-movie script into a vehicle for Lana Turner, he quipped to a friend: "I feel like an obstetrician required to successfully deliver a mastodon from a beaver." Some tasks are simply beyond us.

•

In the downstairs parlor, the one with the portrait of the Scottish ancestor on the wall (my father once shot the canvas with a bow and arrow—if you look you can see the slight pucker where it was repaired), there is a low side table with a number of drawers. Open the one on the right—not the one with bridge decks and score pads, and not the one with the old stereoscopic viewer with slides of the battlefield—but the one that contains twenty-eight daguerreotypes in their tooled leather cases.

Open them, one by one, taking care not to mix up the handwritten slips identifying the sitter, which someone at some point realized had become necessary to include. Captain Edward McPherson in his Civil War uniform (his right hand a little blurry—perhaps he couldn't bear to sit still). A dashing relation named Scott Fletcher with upswept hair and rakish sideburns. A brother of Scott Fletcher (unnamed). A McClellan, a McClanahan, a mother and child, and a stern Presbyterian pastor and his tall wife. A pretty, young girl named Rebekah in a plaid dress. And then there she is again: a severe, skeletal widow wearing a black dress and gloves, her hair drawn tight beneath a white bonnet—a life reduced to the opening and closing of a clasp. One case has a lock of auburn hair pinned to its red velvet lining. The hair is the same color as my dad's—and my own—when we were young boys. No one knows the name of the man in the picture. A friend? A lover? A relative? He has a tidy moustache like a saloonkeeper in an old Western. I joke about sending a strand of the hair for DNA testing.

Some of the images are so pristine you can still see the tinting that was painstakingly applied by hand. So lifelike: apples of rouge on the cheeks and even a little pink on the lips. Other images are degraded, their seals broken, the glass hazy with mold spiders and a sickly green patina that invades the edge of the frame. There are black dots (called "measles") and a delicate white frosting; a sinister condensation bubbles under the glass. The figures are painted in a thin film of mercury on a silver plate as polished as a mirror, which, thanks to a trick of the light, provides a picture of unusual depth. You shouldn't try to clean a daguerreotype, as the image, thin as dust, brushes easily away.

Oliver Wendell Holmes called the daguerreotype the "mirror with a memory." He knew each picture was a one-of-a-kind artifact—it could not be reproduced. Daguerreotypes offered a mirror image; that is, the left and right sides were flipped. The only way to get a true likeness was to make a daguerreotype of a daguerreotype—two wrongs making a right. This was a complex process, so most people settled for a backward portrait.

Edward McPherson took a daily tonic for his health. On December 14, 1895, he reached for the wrong bottle and instead of downing his usual after-dinner drink, he ingested a tincture of nux vomica, an elixir containing a hearty dose of strychnine. The medicine had been prescribed—in minuscule amounts—for an ill member of the family. Edward immediately realized his mistake. A doctor was summoned, but it was too late. Edward died before dawn, leaving a widow and five grown children. McPherson men, it has been said, are nervous around doctors.

Someone once said to me, "There is a fine line between nostalgia and necrophilia." Why this ghoulish obsession with history? Am I really writing about the past, or am I stealing it? And consider the flip side: now, as I type, am I recording the present, or letting it slip away?

Edward lived, wrote, and published before the day of book-jacket blurbs, but I can imagine even he must have smiled in his old age, his bushy beard bristling, to receive a notice from the *New York Times*, which, upon publication of his *A Political History of the United States of America During the Great Rebellion*, declared it "the most complete and perfect record . . . [that] has ever yet been made."

Naturally my wife and I drove our first and only child, a daughter, to Gettysburg. And here is the moment when this essay, this life, takes a turn that you—though perhaps not I—fully saw coming. Our daughter would be baptized in the same church that I was, and my father was, and his father was, and so on. She'd wake in the same house, perform the same morning ritual (a bolted breakfast, some itchy clothes), and head up the same few blocks to the plain Presbyterian church that has a pew where Lincoln sat.

I used to think that having a child is one of the ways we enter into history. But, in the end, what is passed down? What if history weren't a billowing, backward-looking angel but a dark, quiet smudge? Put another way: we stand before our fathers, and their fathers, holding our breath, shocked mute before a chain of fun-house mirrors vanishing in a point.

Our girl is fully comfortable playing in the house of her ancestors. It is late summer, hot. My father is goofy, proud, obsessed—your typical grandpa, maybe, but not my typical father. He wants her—who, at less than a year old, is not yet speaking—to call him "Teddy," a name I never heard anyone call him other than his mother, who had died twelve years before. An improbable name, a boy's name, as if time were slipping backward. He worries about this name from the moment we announce my wife is pregnant. He begins signing it in emails.

We invade the home with an aggressive clutter: large portable crib, diapers and wipes, brightly colored, eco-friendly toys strewn about the hardwood floors, blankets, snugglies, extension cords snaking toward various infant apparatuses, including a sound-machine-slash-night-light sighing and swooshing beneath the crib. Hanging in an otherwise empty closet is a pristine white baptism dress. We feed the baby mushed bananas in the same cracked melamine bowls I used when I was her age. She sleeps soundly in her new room. For the first time, a baby monitor sends invisible frequencies skittering through the old house. The walls are so thick the signal quickly degrades; a few rooms away, the receiver is fitful, crackling with ghosts, a faint ticking, a clock: Loss, Loss, Loss.

Late afternoon: in the middle of the lawn off the side porch, my father splashes with the baby in a blue plastic pool, which he purchased before our arrival and will happily fill up with the garden hose at the drop of a hat. They enjoy getting each other wet; she in a lady-bug bathing suit, and he in khaki shorts and a leather belt. It is a silly scene, cheerfully indecorous, one no doubt mirrored across the country on this bright summer day, but here it still seems improbable, somehow, with those shades of dark-suited ancestors tending flowers and taking tea in the pergola, now fallen, whose footprint still haunts the upper yard.

In the house, the past lurks in numbers, accounts, deeds, portraits, and books—the piled-up dead. Inside, when I work at my desk, I feverishly hope two wrongs can make a right. But outside, this moment: this can never be recorded. This history is hers, not mine, and she is too young to remember. My wife snaps a picture, but it is too late. The hose is off. Already the water is drying on my daughter's skin, which now has grown cold, and behind her my father is turning his back to disappear into the house.

2013

Open Ye Gates!
Swing Wide Ye Portals!

*I hand the attendant a fifty-cent piece and watch him drop it into the automatic turnstile, itself a marvel. Behind me, the murmur of money changers, the swish of gored skirts tapering to white shirtwaists. Beyond that, the din of St. Louis. My sack suit rustles as I stride ahead. The stile makes a quarter turn, an electric pulse registering my attendance in a distant room. Officially, I've arrived, but I am not here. I'm crossing the threshold of an impossible city: a manicured wonderland of symmetrical lagoons winding through sculpted gardens studded with allegorical statues—in the distance, rising like white gods, loom the massive palaces of learning, their beaux-arts façades harkening back to ancient Rome and heralding a future brighter than the hundred thousand incandescent lights that line them against the night. Perspective fails; buildings rise and fall with the logic of a dream. There is music in the shadows. My pulse quivers in my throat. I'm dizzy and jumpy and—underneath it all, for reasons I cannot name—more than a little sad. I have all evening, but time is running out. None of this can last. The words of Exposition President David R. Francis ring in my ears—*Open ye gates! Swing wide ye portals! Enter herein ye sons of man, and behold the achievements of your race! Learn the lesson here taught and gather from it inspiration for still greater accomplishments!—*and I step into the fair.*

St. Louis is a city of gates that do not normally swing wide. The urban private street, or "private place," is believed to be a local invention, dating to the 1850s. Private places are owned by their residents, who typically build and maintain the road, median, sidewalks, curbs, street lighting, and—most crucially—gates. Some gates were utilitarian, imposing and plain; others were small castles, complete with clock towers, fountains, statues, gaslights, and gatehouse apartments that caretakers (and, later, college students) lived in until the 1980s. Private places offered a refuge from the ever-booming city, a world apart. Some have been razed, their gates uprooted, the neighborhoods now troubled by crime; many still stand, pockets of wealth and privilege, with boards of trustees that oversee matters of law (historic preservation, landscaping) and etiquette (street parking, book clubs, Easter-egg hunts).

Nearly two years ago, when my wife and I were moving to town and looking for an apartment, we were taken aback: everywhere, gates, gates, gates. Gates that lock and unlock according to byzantine schedules publicized only to residents (thus thwarting commuters and anyone else who might try to cut through the neighborhood). Gates that open by remote control. Rolling metal gates with yellow hazard signs. Gates built for carriages that now barely fit a car. Even in less rarified neighborhoods—with weeds in the lawns and unwashed economy sedans on the street—at the end of the block might stand a pitiful (and easily dodged) sawhorse made of white PVC pipe. A symbol that speaks to the natives. PRIVATE STREET: NOT THRU. PRIVATE STREET: NO PUBLIC PARKING. NO THRU TRAFFIC. PRIVATE NEIGHBORHOOD. NO SMOKING BEYOND THIS GATE. PRIVATE. NO TRESPASSING. KEEP OUT.

The Louisiana Purchase Exposition, popularly known as the 1904 World's Fair, opened in St. Louis on April 30, one hundred and one years to the day after the signing of the Louisiana Purchase Treaty.

Before a crowd of 187,793 people, John Philip Sousa's band played the "Star-Spangled Banner," five hundred choristers sang the "Hymn of the West," the fair's official song, and—from the East Room of the White House—President Theodore Roosevelt touched the gold telegraph key that sent the signal to unfurl ten thousand flags and begin pumping ninety thousand gallons of water a minute down the three terraced "cascades" that flowed into the Grand Basin, where four fountains threw water seventy-five feet into the air at the foot of Festival Hall, the centerpiece of the fair, a building that boasted—in addition to the world's biggest organ—a gold-leafed dome larger than St. Peter's.

At first glance, the fair offered a spectacle of size, a vision of man's enlightened expansion into—and conquest of—untrammeled space (recalling contemporary notions of the Louisiana Purchase itself). The fairgrounds occupied 1,272 acres—double the size of the famed 1893 Chicago World's Fair—spilling out of the city's giant Forest Park onto the campus of Washington University and neighborhoods to the south. Twenty thousand people would live and work on the grounds. In preparation, crews straightened and buried a river in sixty-five days. They built 1,576 buildings, plus a garbage plant, sewer, post office, press pavilion, telegraph stations, pay telephones, and 125 eateries that could feed 36,650 people at one time. Five of the restaurants could seat more than two thousand people. Visitors ate everything from filet mignon to frankfurters, fried frog legs to caviar, plus international delights such as Japanese sukiyaki, Mexican guacamole, Indian curry, and Egyptian molokhia soup. To drink: 1893 Louis Roederer brut champagne (six bucks a quart), mineral water (sixty-five cents), or Jameson's whiskey (fifteen cents a shot), not to mention—this being St. Louis—plenty of beer. There were five fire-engine houses and thirty-six miles of pipe serving a network of sprinklers and hydrants, some of which to this day still dot Forest Park, popping up incongruously on the golf course.

The great distances between attractions made the fair taxing to navigate. Visitors traveled by intramural railroad, a trolley that trundled twelve miles per hour through seventeen stops; they boated along the mile-and-a-half system of lagoons in gondolas or swan and serpent boats. They rented a rickshaw or wheelchair, with or without a guide to push; drove a car; or, if the mood struck, rode a camel, burro, or giant turtle. The official guide claimed a "brief survey" of the wonders would require a minimum of ten days and fifty dollars.

The colossal exhibit palaces were built of yellow pine and ivory-colored "staff," a mixture of plaster and hemp that could be easily molded, sliced, sanded, and sawed. On average, eighteen trains of forty cars each were needed to haul the materials for a single palace. There were some seventy thousand exhibits from fifty-three foreign countries and forty-three states (plus more than a few territories and businesses). The fair offered a taxonomy of knowledge: exhibits were sorted into sixteen departments that were divided into 144 groups that were subdivided into 807 classes, an encyclopedic education open to all and structured to create, in the words of the director of exhibits, "a properly balanced citizen capable of progress." The goal: to show civilization marching proudly in a direction. The faith: that from the artifacts of the past one could draw a line to the future. In practice, the fair fostered fierce national competition under the banner of international exchange. Proudly on display: progressivism, nationalism, exoticism, racism, hucksterism, humiliation.

The fair's president, David R. Francis—local businessman, Democrat, former St. Louis mayor, the state governor who failed to win the bid to host the 1893 exposition, and now, eleven years later, one of the most photographed men in the country—would proclaim of his fair: "So thoroughly did it represent the world's civilizations, that if all man's other works were by some unspeakable catastrophe blotted out, the records established at this exposition by the assembled nations would afford the necessary standards for the rebuilding of our entire civilization." A time of optimism, these years between the Gilded Age and the First World War.

I wander the palaces, open from nine until dusk. Afterward, I walk the grounds until half an hour before midnight, when the fair lights are gradually, almost imperceptibly dimmed to dark. . . .

The Palace of Electricity is a cathedral of dynamos, motors, rheostats, transformers, and vacuum tubes. I touch the wall—the building hums. Meanwhile, I am speechless before the radiophone—sound transmitted over a beam of light! They are perfecting wireless telegraphy. I fling a message to Kansas City: "Wish you were here." A man offers to show me the power of lightning. His companion says he can record and replay voices on a steel wire. Lights are everywhere—big, small, colorful, and bright. Inventors claim soon our homes will have wall outlets. I ponder the mysteries of electromagnetism, electrochemistry, electrotherapeutics, and electric cooking. A hefty gent clutches his wife: "Steak done in six minutes—lobster broiled in twelve!"

The central court bustles with crowds that circle aimlessly, heads bent. The yard is silent save for small, exultant sighs. A man bumps into me and, with a nod, moves on. He is wearing earpieces that sprout from a curious wheel he holds in one hand. A dusty farmer takes off his hat, then puts it on again—over and over, an incredulous, unconscious salute. An old woman stands on her toes, as if straining to the heavens. A little girl holds her skirt, her mouth hanging open. Someone hands me a wireless receiver; I strap it on—suddenly I hear music pulled from the ether!

In the 1944 MGM musical *Meet Me in St. Louis,* Judy Garland plays a young spitfire trying to snare the boy next door in the months leading up to the 1904 World's Fair. The exposition's unofficial anthem—"Meet Me in St. Louis, Louis," a ditty about a man who returns home to find his sweetheart has fled their humdrum life for the bright lights of the fair—can be heard at least six times in the film's first five minutes. (To this day, the song turns up all over town; at my first hometown baseball game, I was not surprised to be led, on the jumbo screen, in a sing-along by the St. Louis Symphony clad in Cardinal red.) Nominated for four Academy Awards, *Meet Me in St. Louis* contains several classic numbers, among them "The Trolley Song" and "Have Yourself a Merry Little Christmas." A darling of best musical and film lists, it has been named "culturally significant" by the Library of Congress and preserved in the National Film Registry.

The exposition saw almost twenty million visitors during its seven-month run—about one hundred thousand a day. (On weekends, trains to the fairground left downtown's Union Station *on the minute.*) The fair offered an unparalleled economic boon to the city that had lost the chance to host the 1893 Columbian Exposition to its great midwestern rival, Chicago. In 1904, St. Louis was the nation's fourth largest city, centrally located on America's two largest rivers, a rail and river hub that—according to the official fair guide—claimed the biggest brewery, tobacco factory, cracker factory, and chemical-manufacturing plant in the country; the largest brickworks and electric plant on the continent; and one of the grandest shoe operations in the world. The city also churned out hardware, drugs, saddles, white lead, jute bagging, hats, gloves, caskets, and streetcars. Its Union Station was the terminus of twenty-seven rail lines; its citizens read nine daily papers and sent their children to the nation's second-best school system. That said, St. Louis had suffered an economic depression from 1893 to 1897 and weathered a bloody strike of streetcar workers in 1900; the local government was plagued by corruption and graft, the city interests run by a cabal of businessmen called "the Big Cinch." In 1902, *McClure's Magazine* dubbed the city one of America's "worst governed." For St. Louis's new Progressive Reform mayor—busy cleaning up the water, air, streets, and government in time for the grand opening—the fair was a chance at redemption through political force.

Initial funding was raised through equal parts federal appropriation, local municipal bonds, and sale of ten-dollar shares of fair stock to the good people of St. Louis. The exposition was meant to honor the centennial of the Louisiana Purchase of 1803, itself a shrewd deal, the

U.S. government shelling out fifteen million dollars to France for what would become thirteen states west of the Mississippi. Due to delays, the fair missed the anniversary, which gave St. Louis the chance to steal—after threatening to hold its own rival athletic games—the previously scheduled 1904 Olympics from Chicago. The fair would have it all, including sweet revenge.

I stroll the "Model Street," block after perfect block courtesy of the Municipal Improvement Section of the Department of Social Economy. A man loafs on the wide lawn, his collar open, before a guardsman tells him to keep moving. I pass a town hall, a hospital, a civic pride monument, and a playground, where lost children are gathered. By the end of the fair, all 1,166 of them will have found their way home. For a small fee, a woman checks her two-week-old infant with a nurse. She waves: "Mother will be back soon!"

Opened in 1954, Pruitt-Igoe was a mammoth, state-of-the-art public housing development designed by Minoru Yamasaki, who later would build another striking set of modernist towers, the World Trade Center. The Pruitt-Igoe development rose on a parallelogram bounded by Cass Avenue, North Jefferson Avenue, Carr Street, and North Twentieth Street on St. Louis's north side, close to downtown. A city-designated "slum" was razed and in its place rose a modern utopia, Le Corbusier's "Radiant City" made real: thirty-three modular eleven-story buildings smartly arrayed in superblocks across fifty-seven acres, each high-rise facing the same direction, vertical neighborhoods of light and space with ample parking and vibrant public areas. Kids scampered in the breezeways. Apartments were clean and bright, offering views better than those enjoyed by the rich. In a recent documentary, *The Pruitt-Igoe Myth,* a former resident remembers her top-floor apartment as a "poor man's penthouse." Another says, "It was like another world," then adds, "Everybody had a bed."

Pruitt-Igoe was founded on the faith of communal, public life; it offered better living through architecture. One set of high-rises was to be white (the Igoe Apartments, named after a Congressman), the other black (the Pruitt Homes, after a Tuskegee airman), but *Brown v. the Board of Education* came down the year the development opened—and the whites moved out. Black by default, Pruitt-Igoe flourished. In 1957, occupancy was at 91 percent. Fifteen thousand tenants would call it home.

The fair's fanned grounds—laid out by George Kessler, the architect of Kansas City's parks and boulevards—offered a mix of the urban and pastoral. The landscape was strewn with twelve hundred staff statues that, according to the chief of sculpture, aimed "to create a picture of surpassing beauty and to express in the most noble form which human mind and skill can devise, the joy of the American people at the triumphant progress of the principles of liberty westward across the continent of America"—though at least one fairgoer sniffed, "It is a pity that there are so many statues exhibited even on the grounds absolutely naked."

Only twenty-five years old when construction of the fairgrounds began, Forest Park was a wilderness in the process of being uplifted—still more forest than park. In September 1901, President Francis and his party of VIPs were an hour late to ceremonially drive in the first stake because they were lost in the wilds of the park's northwest corner. Then steam shovels moved earth and hills, lakes were drained, and century-old trees felled. Despite the exposition company's contractual obligation—spelled out in 1901 St. Louis Ordinance 20412—to restore the park to its original form within a year of the fair, there was no going back. Forest Park had become a groomed urban oasis, and wrangling between the city and the company lasted for years.

I pay fifty cents and step into the sky, courtesy of the Giant Observation Wheel, the invention of Mr. George Washington Gale Ferris Jr. (now deceased), who envisioned a perfect circle spinning above the plain. I board one of the thirty-six cars at random, but I've made a happy choice. After sixty of us crowd inside, a giddy couple announces they will be married at the top. They're both sitting on ponies. A piano stands in the corner. The guard tells me he's seen it all. Yesterday, a female daredevil made an entire revolution standing atop a car. A few cabins below, fashionable ladies and gents are enjoying a private banquet. The wheel is so quiet we can hear the tinkling piano as we're swung twenty-five stories into the air. While the crowd cheers the kissing couple, I hold my breath—I can see the whole world: the Grand Trianon of Versailles, Charlottenburg Castle, the Orangery at Kensington Palace, a Roman villa, a Chinese summer palace, Robert Burns's Cottage, and the homes of Andrew Jackson, Jefferson Davis, and Thomas Jefferson—all of them rebuilt at the fair. A city of replicas, a cosmopolitan capital forged of iron will.

Today, the city's most conspicuous monument to Thomas Jefferson and the Louisiana Purchase is the massive Gateway Arch, designed in 1947 by Finnish futurist Eero Saarinen, who died before he could see the arch dedicated in 1968 in honor of America's westward expansion. Arches were popularized but not invented by the Romans. There are many kinds of arches: horseshoe, lancet, scheme, ogee, trefoil, basket-handle, Gothic, Tudor, triumphal. Saarinen's monument was a six-hundred-and-thirty-foot-tall-and-wide stainless-steel curve made of tapering equilateral triangles, a mathematical dream rising over the heartland.

The lore of the fair claims many firsts: the debut of Dr. Pepper, the ice cream cone (known as "World's Fair Cornucopias"), iced tea, hotdogs, hamburgers, cotton candy (a.k.a. "Fairy Floss")—but these items were merely popularized and not, as legend might have it, invented at the fair. There were several true firsts: the first appearance of puffed rice cereal, which the Quaker Oats Company shot out of eight cannons every fifty minutes; the first large-scale cast of Rodin's *Thinker*; the first participation from China in an exposition; the first Japanese garden in America; the first time British troops paraded on U.S. soil since the Revolution.

Perhaps foremost: the first Olympic Games played in the U.S., which also saw the first gold, silver, and bronze medals handed out, and took place in the first concrete and steel stadium, Washington University's Francis Field, which had room for fifteen thousand. Competitors from the U.S. and eleven foreign nations set thirteen Olympic records in twenty-two official events.

Notable performances included a gymnast, George Eyser, whose wooden leg didn't prevent him from winning six medals, three of them gold, and an unsportsmanlike brawl after the fifty-yard swim. But the most memorable event was the marathon, which was run August 30th at three o'clock in the afternoon in ninety-degree heat over tough terrain and dusty roads. Runners received only two chances for fresh water—at six and twelve miles—in deference to the head of the Department of Physical Culture's amateur scientific interest in dehydration. Fewer than half of the thirty-two participants crossed the finish line. Runners were plagued by traffic, hills, and cheating. (The first man to return to the stadium received a wreath from Alice Roosevelt, daughter of the president, before it was revealed he had ridden eleven miles in a car.) The true victor, Thomas Hicks, pride of the Cambridge YMCA, ran a time of 3:28, though aided by brandy, raw eggs, and the stimulant/rat poison strychnine. After being sponged by his supporters with hot water from a car radiator, he had to be carried, hallucinating and shuffling his feet in the air, across the finish line. He had lost eight pounds. Men representing clubs from Chicago and New York stumbled in second and third. A Cuban mailman came in fourth (he had hitchhiked to the games). Len Tau, a Tswana tribesman running barefoot, finished ninth and presumably would have done better if not driven a mile off course by a wild dog. By taking time off from the Boer War exhibition, he and his fellow countryman, Jan Mashiani, who finished twelfth, became the first South African Olympians. Black athletes wouldn't represent the country again until the 1992 games.

Education was the theme of the fair—which was meant to be "an international university" concerned not with commerce but knowledge—

but not all exhibits were meant to uplift. More liberal entertainment could be found on the mile-long midway called the Pike, where battle reenactments, hootchy-kootchy girls, ragtime rhythms, and flights of wild fancy thrived outside the purview of the Bureau of Music and the Department of Art. The Old Plantation featured log cabins and cakewalking "slaves." The Jerusalem re-creation was said to include one thousand natives of the city, though one magazine reporter found a fellow from Hoboken. Battle Abbey included cycloramas of the battles of Gettysburg, New Orleans, and Manila, plus Custer's Last Stand. Jim Key, the educated horse, could spell and sort mail. He was not the only equine wonder; in the Boer War reenactment, even the horses played dead.

On the Pike, one visitor observed, "No respect was shown to age or dignity, no mercy to starch and feathers." Bands of dancing young men might accost couples, and "every stiff hat was a target for the inflated bladder" (or water balloon). The same fairgoer wrote in his memoir, "I believe if the Pike had been a mile longer it would have led to hell."

Later, he recanted: "And yet I had a desire to imbibe a little of the spirit of the Pike. I wanted to be a boy again. Be a little bit bad perhaps."

The Palace of Agriculture is a blinding colossus in the sun. The man next to me reads from a booklet: twenty acres large, covered with 147,250 panes of glass. I have timed my visit—in one minute a giant clock made of thirteen thousand flowers will strike noon. I am finished with the exhibits. I have seen the Missouri corn palace, the 4,700-pound cheese; I have laughed at Minnesota's contribution, "The Discovery of St. Anthony Falls by Father Hennepin" shaped out of 1,000 pounds of butter. Now a hiss of compressed air throws the 2,500-pound minute hand the final 5 feet, where it points to the giant numeral 12. An hourglass flips, doors open to reveal the gears of the clock—oh, triumph of industrial time!—and a massive bell tolls the death of more agrarian rhythms.

The company is a major employer in this city. One cannot miss its print and radio campaign: "We grow ideas here." "We work together here." "We dream here." "We're proud to be St. Louis Grown." Its website offers videos of employees working in food banks, cleaning up after tornados, visiting Forest Park, and standing in front of the arch. Articles rate the town's best burger joints, as judged by company workers. The company is a major donor to local charities and institutions, including the university in which I teach. In 2013, the company's net sales were $14.8 billion, up 10 percent. Its chief technology officer won the 2013 World Food Prize. The company has 21,183 employees in 404 facilities in 66 countries—but its headquarters are here, where, over the years, the much-maligned Monsanto Company has worked to produce saccharin, PCBs, polystyrene, DDT, Agent Orange, nuclear weapons, dioxin, RoundUp, bovine growth hormone, and genetically modified seeds.

Pyramids of fruit on a sea of china plates—the entire Palace of Horticulture smells like apples. Virginia has created a statewide shortage by sending too many to the fair. Part of me is sad I missed Missouri Peach Day, when the palace gave away fifty thousand peaches. I dip my fingers into the fountain, which gushes ice water. Farmers shake their heads at the monstrosities on display: a pineapple the size of a turkey, a mysterious dimpled fruit that is said to be the unholy cross between a strawberry and a raspberry.

I live in a small apartment building that stands in the footprint of the Horticulture Palace. We grow nothing in the backyard but herbs, potted lettuce, and a few stunted rose bushes, but on sunny days I like to think I smell apples.

In *Meet Me in St. Louis,* Judy Garland's older sister reminds her, "Nice girls don't let men kiss them until after they're engaged. Men don't want the bloom rubbed off"—to which Garland responds, "Personally, I think I have too much bloom." Garland gives a spirited performance as a virginal teen, her eyes flashing beneath a swath of auburn hair coiled on her forehead like a fender, but in reality the twenty-one-year-old ingénue already had been persuaded to have an abortion and would soon move in with the director, Vincente Minnelli, nearly two decades her senior. (They married a year after the film came out, had a daughter—Liza Minnelli—and divorced five years later.)

The film's crisis comes when Garland's father declares his intention to move the family to New York City, dashing his daughters' romantic interests and hopes of the fair. In the dramatic Christmas Eve denouement, he decides the family will stay put, saying, "New York doesn't have a copyright on opportunity. Why, St. Louis is headed for a boom that'll make your head swim."

A slick-haired fellow shouts from an automobile: "One hundred and forty models of cars powered by gas, electricity, and steam!" His eyes shine with belief in the Palace of Transportation. He waves a magazine furiously about; as he reads, he stabs his finger in the air: "This new form of carriage will become perfected, and then the great cities will spread out into the suburbs, and life on an acre will become a possibility for even the humblest class of people!"

The Pruitt-Igoe housing project was built for a postwar boom that never came. The city of St. Louis was dying; another kind of planned community was thriving beyond its borders—the suburb, fueled by the same 1949 Federal Housing Act that enabled Pruitt-Igoe. Having already legally fixed its boundaries, the city couldn't abate its population decline by annexation. With the middle class fleeing to the suburbs, the development would never be able to raise the significant maintenance fees it needed from its tenants. The city let the brand-new buildings deteriorate almost from the start. More pernicious factors were also at work. Suburbs passed zoning laws barring low-income housing; public projects became a tool of segregation, the goal being to prevent, in the parlance of the day, "negro deconcentration."

St. Louis is often ranked as one of the country's most segregated cities based on what's called its "dissimilarity score," which analyzes racial makeup across census tracts. While different measurements suggest the divide might not be so stark, the traditional color line is widely acknowledged to be Delmar Boulevard. Seventy-three percent of residents south of the boulevard are white; head north, and neighborhoods become 98 percent black. A color-coded map of the 2010 census reveals a similar pattern: blacks to the north and whites to the south and west, with some intermingling along the southeastern edge of the city.

The wealthy west county suburbs are predominately white. One historian has called St. Louis "the poster child of white flight."

In the musical, Garland sings about her home at 5135 Kensington Avenue, a stately three-story Victorian on an idyllic block on the MGM back lot known as "St. Louis Street" (which would appear in a number of films before being torn down). The real address—a few blocks north of Delmar—belonged to the writer Sally Benson, from whose memoirs—serialized in the *New Yorker* as "5135 Kensington"—the film was drawn. Benson's former home was abandoned and demolished in 1994. Today, 5135 Kensington Avenue is a vacant plot with a history of debris and graffiti complaints, valued in the city's last appraisal at $3,800. The St. Louis Land Reutilization Authority owns it. The neighborhood has suffered over the years; in 2001, wild dogs ate a ten-year-old boy in a park two blocks away. ("They fed off of him," the police chief said.) As it turns out, Benson's family *did* move to New York and never saw the fair.

I push past the crowds into the Palace of Education and Social Economy. In a model classroom, the teacher struggles to keep the attention of the giggling local children, who are thrilled at their turn to take part in the exhibit. I wander into the "School for Defectives." Deaf students sewing—blind students on violins! Helen Keller, a senior at Radcliffe, will be lecturing soon. On my way out, I peruse modern treatments for the insane.

St. Louis is a city preoccupied with school districts, perhaps because its public schools were stripped of accreditation by the State Board of Education in 2007, when governance of the district was transferred to an administrative board appointed by the governor, mayor, and president of the city's Board of Aldermen. The district was nearly $25 million in debt. Fewer than one in five students could read at grade level. In fall of 2012, the schools regained provisional accreditation, though the previous spring's exams had shown only 27 percent of students passing in math (compared to a 55 percent average statewide).

The problem of St. Louis schools is a Gordian knot of politics and passion that has been studied by heads far smarter than mine. I cannot do it justice here. But most people admit the usual "solutions"—bussing, lotteries, a district transfer system, a mix of private, parochial, charter, and magnet schools—have failed to create equal opportunity.

The odds are against the shrinking city. In 1970, St. Louis public schools enrolled 111,233 students. In 2011, average daily attendance was 20,880. A 1972 discrimination and desegregation lawsuit lingered until 1999, when a settlement finally ended court supervision of the district. Left in place: a voluntary transfer system that allows African American students from the city to attend one of the participating (wealthier) county districts. Waiting lists are long; each year, thousands of students are turned away. Those who are admitted face an average one-way bus ride of fifty-four minutes, among other challenges. At the start of 2013, five thousand and thirty-six students transferred from city to county schools. The program also allows white suburban students to transfer to city magnet schools—eighty-seven students took advantage of that opportunity.

The transfer program is scheduled to extend until 2019.

East St. Louis, Illinois, sits just across the Mississippi River from downtown St. Louis. Jackie Joyner-Kersee, Miles Davis, Jimmy Connors, and Ike and Tina Turner have called it home. The U.S. attorney for the district recently called it "the Wild West." From 2008 to 2011, the city had to cut its police force by 33 percent; the per capita murder rate is more than sixteen times the national average. Since 1960, East St. Louis has lost two-thirds of its population; a casino and the school district provide most of the jobs. At a party, I met a man who moved to St. Louis six years ago; he told me, "When friends visit from Europe, I drive them through East St. Louis. It really shakes up their notion that there is a certain level of poverty here that no one falls below. They can't believe America can be so bombed out."

A man in a blue uniform approaches me. He is a physical specimen—between twenty-one and forty years old, at least five-foot-eight, 145 to 180 pounds, and of "good feature and bearing"—one of the Jefferson Guard, the omnipresent guides and official policemen of the fair, some three hundred strong. I nod and say, "Just passing through." He looks at me and smiles, but I'm starting to feel uncomfortable, nervous. I cross the street. He pulls his mustache as he reprimands a boy for spitting: "No one is innocent." He should know. For fifty dollars a month plus housing, he and his brethren will arrest 1,439 citizens, including 312 trespassers, 421 disturbers of the peace, 5 murderers, and 1 vile soul charged with "wife abandonment." On hot summer days, he sports a lighter khaki.

As a kid, I learned of East St. Louis from National Lampoon's *Vacation* (1983), in which the hapless Griswold family gets lost there, a sequence that—by playing into negative stereotypes—popularized the city as national shorthand for "crime-ridden ghetto." (A clueless Chevy Chase lectures his family, "This is a part of America we never get to see. . . . We can't close our eyes to the plight of the cities. Kids, are you noticing all this plight?")

But rewatching the movie, one notices that—thanks to a continuity error—the family actually would have been in St. Louis, having already been shown crossing the river into Missouri. In terms of plight, St. Louis holds its own. In 2013, it ranked as the third most violent city in the U.S. after New Orleans and Detroit. Another report listed two of its neighborhoods on the "Top 25 Most Dangerous Neighborhoods in America." Invariably the mayor's office and the police department rebut such rankings—often with good reason, as an antiquated city/county divide puts St. Louis at a tremendous national disadvantage in polling methodology, and skews many of these stats. Still, at the time of the fair, St. Louis was nearly twice the size it is now. It has lost some five hundred thousand people over the past fifty years. Segregation, disastrous "urban renewal" projects, "white flight" to the suburbs, "redlining" (racist lending practices), and blockbusting (racist real-estate scaremongering and profiteering) tell some of the complicated story. The fact remains: the last time the city had this few people was in 1870—and the national perception endures that it is dangerous to live in St. Louis on either side of the river.

Pruitt-Igoe residents were treated with suspicion and subject to dehumanizing regulations, the subtext being that the poor were in need of forced moral uplift. Televisions were forbidden; apartments could be painted no color but white. Disastrous welfare laws broke up families—no able-bodied man could live in a unit that received federal aid—so fathers hid in closets when they were supposed to be, by regulation, out of the state. Many of the buildings' modern innovations functioned poorly. "Skip-stop" elevators that didn't land on every floor—an economic concession that supposedly encouraged mingling and use of the stairs—made residents easy prey for muggers. Public galleries became gauntlets. Residents had been promised beautiful, safe, affordable housing, but city maintenance deteriorated. Elevators smelled of urine and broke down regularly; "vandal proof" light fixtures stayed dark. Firefighters, police officers, and ambulances stopped showing up after frustrated tenants dropped bottles and bricks on them. Pruitt-Igoe quickly became an emblem of an overblown white fear of black poverty and crime. As the experiment unraveled, a complex story of structural inequality and misunderstood urban forces was turned—by some—into a more vicious parable of how *those people* just couldn't be helped, just couldn't be trusted with nice things. Critics blamed the residents, the design, and the welfare state; such accounts overlooked or underplayed the effects of racism, segregation, and a city in crisis. In 1965, *Architectural Forum* noted, "Pruitt-Igoe also is a state of mind. Its notoriety, even among those who live there, has long since outstripped the facts." Rents were raised three times in one year. Residents were stretched to the breaking point and got nothing in return. Apartments went without heat. Buildings fell apart. The city had become a predatory "slum lord."

I hurry past an anthracite coal mine belching soot and smoke in a gulch south of the palace, pausing only briefly to wonder what might be unearthed. Slipping between a pair of Egyptian obelisks, I enter the Palace of Mines and Metallurgy. I glance over my shoulder—no policemen in sight. I have no time for an oil rig, a 1,200-pound pot of mercury, the devil made of sulfur. I duck into a dark room that holds a luminous collection of radium ores. A child presses my hand and whispers, "They look like fireflies." Next door, at 2:30, the U.S. Government will stage a demonstration of the mysterious element. Cosmopolitan *implores I not miss this epochal discovery: "Revealing an energy so powerful, inexhaustible, and apparently so abundant in nature, that its substitution for other forms of light and power now in general use is within the range of possibility."*

A local sociologist named Lisa Martino-Taylor recently uncovered a secret army experiment conducted in the 1950s and 1960s in which St. Louis citizens were unknowingly sprayed with a mixture of zinc cadmium sulfide that she believes might have been radioactive. Chemical sprayers were attached to buildings, schools, and station wagons. Residents were told smoke screens were being tested that might conceal the city from the Russians. In fact, St. Louis was believed to be a fair model of certain Soviet military targets—the perfect Cold War replica. The army was studying fallout patterns. Poor minority communities were targeted. Roughly three-quarters of the inhabitants of the test area—what army documents called "a densely populated slum district"—were black. Cadmium was known to be toxic; the finely ground chemical was easily inhaled. Martino-Taylor has collected stories of cancer. The army admits the aerosol was fluorescent, but it won't say whether it was radioactive. Many documents remain classified. Most of the spraying was done around the Pruitt-Igoe complex, home at the time to about ten thousand low-income residents, some 70 percent of whom were children.

The Baby Incubators didn't hold the only newborns at the fair. The first child born on the grounds was to one of the workers in 1902; Louisiana Purchase (Louise) O'Leary died in 2003, nearly 101 years old. On July 1, 1904, a St. Louis policeman brought an abandoned preemie to the incubator exhibit; on the last day of the fair, the policeman and his wife took the baby home and named her Frances, after the exposition president. She forever would be their "World's Fair souvenir." Meanwhile, an Inuit child was born into far more humble housing, only to die shortly thereafter from the summer heat.

In 1969, Pruitt-Igoe tenants organized a rent strike, a shocking development in public housing. After nine months, the housing authority caved. But that winter, two months after the victory, water and sewage pipes burst, perhaps partly due to an estimated ten thousand broken windows. Sheets of ice cascaded down the façades. Buildings were vacated, then stripped by thieves. Superblocks became ghost towns, the darkened shells offering a high-rise vantage for drug lords looking to evade enemies and cops.

In 1972, the city capitulated—the first three buildings were imploded. The demolition of building C-15 on April 21, 1972, was nationally televised, the spectacular footage spreading so widely that Charles Jencks, the architectural theorist, proclaimed this first stage of demolitions to be the day "modern architecture died." The final eight hundred tenants were relocated, and by 1976 the development had been erased, leaving a fifty-seven-acre scar across the north side.

In the Hall of Anthropology, housed in a university building I can see from my office window, visitors could get tested anthropometrically—their bodies weighed, their skulls, foreheads, ears, and jaws measured—or gawk at a Brazilian shrunken head. On display in the ethnological exhibits: a family of nine Ainus from Japan (supposedly the world's hairiest people), several Patagonian "giants," pygmies from Africa, representatives from more than twenty native North American tribes, and "many other strange people, all housed in their peculiar dwellings, such as the wigwam, tepee, earth-lodge, toldo, or tent." The Department of Anthropology hoped to show "how the other half lives, and thereby to promote not only knowledge but also peace and good will among the nations." Cultural and political imperialism were given a scientific gloss; the virtues of white, western assimilation were roundly praised. In his diary, one fairgoer favored the Ainus, whom he found "not as dirty nor nearly as lazy-looking as the Patagonians."

There are ethical concerns of looking, and I am reluctant to gawk. YouTube videos titled "the worst ghetto in America" scroll through Google Street View shots scored with either a lugubrious sound track or high-octane rap. On a video for north St. Louis, the comments run the gamut—boastful, racist, reasoned, sad—from trolls, residents, rubberneckers, and the like. One well-designed website tracks the deterioration of a single north-side block, comparing a dense map of dozens of houses and stores at the turn of the twentieth century to a diagram of the six buildings that remain—on either side of the street—nearly a hundred years later. Today there is but one house standing. These are old buildings, often made of St. Louis brick, a fine, beautiful brick that was once the pride of the city. In the fourth ward, there is a long-running scam in which thieves set fire to a vacant house; after firefighters come and scour off the mortar with their hoses, the thieves return and cart off the cleaned bricks. Thanks to a policy of demolition and clearance, an inner-city prairie has sprouted, a startling sight. Satellite imagery shows swaths of city blocks turned into gridded green plots of land, with scrub brush, a few trees, some crumbling structures, maybe a house with no roof. A four-way intersection in the middle of the city might look like a forgotten rural byway.

But perhaps I shouldn't take you on these streets. Perhaps I shouldn't be going there myself just to look. A white professor who rents an apartment in a neighborhood in which he could not afford to buy a house—what gives me the right? I don't know these communities. I can't see them for real. Even writing this, I worry I'm not illuminating a problem but feeding a self-perpetuating fear.

The gem of the anthropological offerings was the Philippine exhibit, dedicated to the islands that had become a U.S. territory following the recent Spanish–American War. After walking a bridge across Arrow Head Lake, visitors entered the forty-seven-acre encampment through a reproduction of the walled city of Manila. Ads touted "40 different tribes, 6 Philippine villages, 70,000 exhibits, 130 buildings, 725 native soldiers—better than a trip through the Philippine Islands!"

The most popular—and controversial—part of the exhibit was the Igorot Village. In their loincloths, the tribesmen looked "like bronze statues," according to one female viewer. (A male visitor noted they "seemed to have a tremendous attraction for the ladies.") Secretary of War William Howard Taft wanted the tribesmen in trousers, but the fair's Board of Lady Managers overruled, upholding their idea of science over prudishness. The loincloths carried the day, even late into the year, when the huts were warmed to accommodate the light dress.

Even more sensational were the regular dog feasts, an occasional tribal tradition greatly played up for the fair. Fueled by reports in the papers, visitors brought dogs to the village to donate, sell, or trade; some sources claim twenty canines a week were provided by a local pound, though the number seems apocryphal.

At the Philippine Model School, Igorot schoolchildren serenaded President and Mrs. Roosevelt with "My Country 'Tis of Thee." The President remarked, "It is wonderful, such advancement and in so short a time!" One Igorot chief insisted a telephone be installed in his imperial hut. A photo caption from *Cosmopolitan* that September: "Miss Roosevelt and her friends are amused at the manners and customs of the Filipinos." Four white-hatted white ladies holding small bouquets peer around some shrubbery and laugh, flashing their ivory teeth.

The area occupied by the village is now a tony—gated—neighborhood I sometimes walk through on the way to campus. In the morning and late afternoon, the wide streets are eerily quiet, deserted, the only activity the comings and goings of uniformed domestic help or buzzing groups of gardeners. In a year of wandering through the neighborhood I don't think I've ever seen an inhabitant of one of these houses. I walk past a public school whose mascot was the Igorot until 1974.

A local sixteen-year-old named T. S. Eliot frequented the fair. I visit the reading room of the Missouri History Museum's Library and Research Center, just south of the Life-Saving Lake (now gone), where shipwrecked sailors were rescued daily at 2:00 p.m. I sit beneath a Byzantine dome in the former sanctuary of a synagogue. The room is a whispering gallery: I hear voices, snatches of conversations, echoes where there is no one. Across the room, an archivist laughs, and it sounds like she's behind my ear. I pick up Stockholder's Coupon Ticket #1313, signed "Thos. S. Eliot" and bearing a photo of the boy poet, who gazes slyly down to the left, as if he knows better than to look. He wears a coat and tie, with tidy hair and a tight collar. His alabaster skin is wan and washed-out, and his heavy-lidded eyes are sunken, blurred, and unfocused, almost blind. One large ear is turned, as if listening. He sports a thin, coy smile. The archivist handed me the photo with a shudder. "Creepy," she said. The boy looks like a seer.

The slim, salmon-colored booklet held detachable coupons good for admission any day but Sunday, when the fair was closed. Tickets were nontransferable; in a fifty-coupon book, only one remains. The cardboard cover is a bit warped, perhaps by sweat. How many hours did it spend inside a pant or coat pocket—or clutched in a sweaty sixteen-year-old palm? The ticket feels heavy with history. Despite being an official stakeholder, the poet never wrote directly of visiting the fair—though a year later, in his school's journal, he would publish a south-seas short story critical of the powers of civilization, certain details of which recall the Igorots. The white boy fell under the sway of "the exotic." Lecturing at Harvard twenty-seven years later, he would state, "Poetry begins, I dare say, with a savage beating a drum in a jungle."

Also on display was an Mbuti pygmy named Ota Benga, who for a nickel would bare his pointy filed teeth. In two years, he would be exhibited in the monkey house of the Bronx Zoo—causing a furor. He eventually relocated to Lynchburg, Virginia, where he went by the name Otto Bingo. In 1916, after filing off the caps on his teeth, he lit a ceremonial fire and shot himself in the heart. There is no poetry in a bullet to the heart.

I distrust my eyes. Chief Geronimo sits in a booth at the Indian Building. A sign says the seventy-five-year-old Apache prisoner of war arrived from Fort Sill, Oklahoma, under military guard. A man whispers, "Yesterday, he made a bow and arrow for my neighbor." When the chief doesn't move, the guy shuffles away. I say, "Chief, I am sorry. I am repulsed by your treatment." He sighs and says, "Go ahead—I know you want my picture." I'm silent. He looks disappointed. "Two dollars, white man." As I reach for my Kodak, I ask, "Is it true the Indians in the Cliff Dweller attraction aren't Hopis, but Pueblos—and they're faking both the snakes and the steps in their 'Hopi Snake Dance?'" He says, "I have nothing for the likes of you."

The Japanese exhibit was a popular but befuddling entry, one that threatened to derail the fair's anthropological narrative. The Ainu, with their light skin and "Caucasoid appearance," provided a problematic example to a contemporary anthropologist, who noted in confusion, "Here we find a white race that has . . . proved inferior in life's battle to the more active, energetic, progressive yellow people, with which it has come into contact." Never mind that Japan was busy beating Russia in a war. A fairgoer wrote in his diary: "Japanese section—wilderness of pottery—clouds of fine needle work. Their every work marvelous." A writer for *Cosmopolitan* was caught short: "First surprise; then profound astonishment; then mortification: this describes the feelings which developed as I made my progress through the Exposition, everywhere Germany and Japan displaying a superiority for which, I confess, I was in no way prepared."

The same didn't hold true for the Chinese exhibit for at least one visitor, who wrote in his memoir: "Cared little for the China section. Somehow it always impressed you as topsy-turvydom, and could not get your mind down to a patient scrutiny, if general, of twisted dragons, fantastic dogs, full-bellied figures, and a confused mass of carvings, jim-cracks, and—and space occupiers that center the thought mainly on the time that it took to make these things." When the fair ended, twenty-nine-year-old Prince Pu Lun, who had lived at the George Washington Hotel for two months, gave the entire Chinese Pavilion to Exposition President Francis, after which it went missing.

On the north side of St. Louis stand the city's many chop suey houses. Happy Chop Suey, Wing Hing Chop Suey, Harold's Chop Suey, Mandarin Inn, Newstead Chop Suey Catfish. There are bars on the windows and no tables or chairs; orders are taken behind bulletproof glass. Chop suey—Chinese vegetables and meat stir-fried in a heavy cornstarch sauce—is a nineteenth-century immigrant invention born in America and embraced early by African Americans. (February 26, 1926: Louis Armstrong records his hit "Cornet Chop Suey.") The country experienced a sustained chop suey craze. But when Nixon went to China in 1972, he didn't eat chop suey. America learned about other—perhaps more authentic—Chinese cuisines. Today, chop suey endures only in Middle America—and in St. Louis, chop suey shops cling sharply to the color line. South of Delmar, they are for the most part simply "Chinese restaurants."

The specialty of St. Louis chop suey is something called (curiously) the St. Paul Sandwich, a gastronomical gut-bomb of mysterious origin. To outsiders, the ingredients might seem a bit topsy-turvy: egg foo young, tomato, pickle, lettuce, and mayonnaise slapped between white bread.

Hungry, I stop at Dragon Chop Suey on Kingshighway because the neon sign says it is open. The menu takes up most of the wall: a plain St. Paul Sandwich costs $2.30 ($3.20 for duck in it, $3.50 for tripe). Extra cheese is 60 cents. Packets of soy sauce, sweet and sour sauce, and red pepper are a nickel each. A yellow sign declares NO CHANGE. There is only one opening in the wall, a rotating octagon made of very thick plastic—one of those distrustful lazy Susans that only lets one side (the customer or the kitchen) access it at a time. The place is deserted. I bend and call into the opening, "I'd like a St. Paul Sandwich." No reply. I've made a mistake in coming here. I try again, but I'm ignorant—ten thirty in the morning is too early for a St. Paul Sandwich. A face appears in the octagon and simply says, "Eleven o'clock."

I've only lived here a year and a half—I don't pretend to know these streets. I am new at the fair, a yokel trying to take it all in. I offer no answers; I might have the wrong questions. This essay is an exposition, its own kind of fair. I am putting together something that doesn't exist. A replica. Perhaps an ideal. My own not-so-radiant city. I am building on sand and shadows; I, too, am destined to fail.

Later, during the lunch rush at Yan Wu House Chop Suey on Delmar, a St. Paul Sandwich takes fewer than five minutes to make. You don't have to be patient; there's nothing to scrutinize. The sandwich costs the seemingly standard $2.30 and is hot and crisp.

August 12 and 13 were Anthropology Days, a series of sporting contests organized by the departments of Anthropology and Physical Culture a few weeks before the Olympic Games (when pervading beliefs held that the Americans would win over the "primitive" races, never mind the fact that George Poage, an African American sponsored by the Milwaukee Athletic Club, would become the first black medalist when he won bronze in both the 220- and 440-yard hurdles).

And so the Sioux competed against the Arapahos in tug-of-war. Tribesmen tossed a fifty-six-pound weight. There was a mud fight. Crack spear-throwers struggled with the javelin. Runners stopped and ducked under the finish-line tape. Participants laughed at the events; they didn't try very hard. The white man's games didn't translate. Attendance was poor, as was the quality of "data" collected by the departments. Winners were given American flags. A Filipino Negrito named Basilio was the fastest to climb the greased pole.

On July 2, 1917, near Fourth and Broadway in East St. Louis, a black man was cornered and strung up on a telephone pole; when the rope broke, the man fell to the gutter, where, according to the *New York Times,* a mob "riddled his body with bullets" before hanging him again. In 1916 and 1917, some ten thousand African Americans moved to East St. Louis from the rural south as part of the Great Migration, greatly feeding white cultural, economic, and political fear. Labor tensions ran high. The night before that July lynching—which would prove to be only one of many—a car of white men had driven through Market Street, shooting at black residents. When plainclothes police officers appeared in a car, they were mistaken for the original culprits and were fired upon, killing two. East St. Louis exploded into some of the bloodiest race riots in American history. The goal: drive out the blacks.

Houses were set alight and the fleeing residents gunned down. Eyewitnesses described babies being tossed into the river or shot in the head and fed into the flames. Small boys fired revolvers. Two young white girls dragged a black woman off a streetcar; another white girl stomped on a black man's face, bloodying her stockings. Bodies were left in the street. Militiamen were ordered to shoot to kill in their efforts to subdue the white mobs, but one black woman—hearing gunfire—fled an outhouse only to have her arm shot off by a soldier. The city's most famous expat, Josephine Baker, would remember watching—as an eleven-year-old—a man being beaten, and hearing about a pregnant neighbor whose baby was torn out.

The mayor's office attempted a cover-up; his private secretary ordered police and militiamen to smash cameras and arrest anyone attempting to photograph the violence. But the next day, a telegram reached the War Department: "Very bad night fires and rumors period. A lot of negroes killed number unknown period." The approximate death toll: eight whites and anywhere from forty to hundreds of blacks. At least $400,000 in property lost. More than six thousand African Americans would flee. W. E. B. Du Bois, sent to bear witness to the massacre, reported an old woman's lament: "We can't live in the South and they don't want us in the North. Where are we to go?"

In the fall of 2013, an editorial appeared in the *St. Louis Post-Dispatch* proposing an amendment to the state constitution that would join the city and county of St. Louis—undoing the "Great Divorce of 1876," what the editorial board called "the biggest mistake this region ever made." Secret talks—including "key city, county, civic, and corporate leaders"—had been underway for years. The paper pushed the mayor to go public. An anonymous opposition group set up a website. Weeks went by with no more news. A longtime St. Louisan told me, "Oh, there are rumors all the time."

I'm having lunch with a prominent local, someone who—along with Maya Angelou, Chuck Berry, Yogi Berra, Phyllis Diller, Robert Duvall, Walker Evans, Redd Foxx, Betty Grable, William Holden, and Stan Musial—has a star on the St. Louis Walk of Fame. We discuss a city that doesn't know how to define itself. During the Civil War, it had slaves but was not southern. It is not western, like Kansas City, on the other side of the state, but it is not eastern, not really—just ask any transplant from the coast. It truly is Middle American, whatever that means. There's the old joke: what kind of city would advertise itself as a jumping-off point, an exit door, a gateway to somewhere else?

It is a city that lives with a sense of belatedness. A city haunted, proud of its past but worried its best days are behind it. The fear: that it has become just another link on the Rust Belt, the next Detroit. It once competed to be the biggest and best in the middle of the country—a crown long since lost to Chicago, though the rivalry (which sometimes seems a little one-sided) endures. It is a city preoccupied with the order of things. Living here, one of the common questions is "Where did you go to school?"—which my lunch companion tells me means high school. In other words, it's a way to talk implicitly about neighborhood, class, religion, and race—an arcane social code that no transplant will ever fully crack. My companion says that even after living here some thirty years he feels like an outsider.

Valentine's Day weekend 2014 marked the 250th anniversary of the founding of St. Louis, which was celebrated with a lavish festival in Forest Park. (The centerpiece: a sculpture of a heart—on fire—rising above the Grand Basin.) The semiquincentennial celebration lasts all year, with exhibits, symposiums, and historical reenactments. For the planners, optimism runs high.

In 2005, St. Louis adopted a comprehensive plan for the use of every block in the city. It has been amended ten times. Supposedly, a north-side renewal is afoot. Land has been bought, the developers given massive tax incentives that were temporarily blocked by lawsuits.

Many north-siders feel alienated, not included in the plan. The biggest developer—whose website details his vision for a fifteen-hundred-acre site that would become "a shining example of modern America" and "a gateway for liberty"—has quoted Winston Churchill: "You'll never get to your destination if you worry about every barking dog along the way."

Today apartments are available (a banner: "Downtown Living, Uptown Style!"). A local nonprofit organization held an international contest to reimagine use for the Pruitt-Igoe site, with winners announced in 2012, the fortieth anniversary of the demolition. The top design, from two Harvard graduate students, took home $1,000 and called for the abandoned land to become an "ecological assembly line," with nurseries and aquaculture basins producing native plants, trees, mussels, and fish. The proposal is beautiful, part memorial, part farm, but I cannot stop staring at images of the implosion. Something about them arrests me. Eventually, I see it. Behind the buildings and the dust cloud, to the left of the horizon, stands what was then the city's most recent monument to a radiant future—dedicated only four years earlier—the gleaming Gateway Arch.

From the beginning, the purity of the plan was suspect. After Saarinen announced his design, an Italian architect claimed the idea was his, stolen from a fascist monument he had drawn up for Rome's 1942 World's Fair (never realized). Saarinen said his arch was universal. (Indeed, Mussolini's arch—while meant to be taller—looks eerily identical.) Twenty-one years later, at the dedication, Vice President Hubert Humphrey proclaimed the Gateway Arch would provide a "new sense of urgency to wipe out every slum," promising that—by its example—"whatever is shoddy, whatever is ugly, whatever is waste, whatever is false, will be measured and condemned."

Forty downtown blocks were leveled to make room for the arch, many of them home to poor bohemians, artists, and African Americans. A day or two after reading Humphrey's speech, I hear a historian say on the radio that the arch hasn't transformed the city as its builders had hoped, and if it is destined to be remembered by history, it will not be as a celebration of the Louisiana Purchase, but as a monument to the mid-twentieth century—to an America so powerful, so brash, so sure of its future that it would destroy a downtown to put up a symbol.

Reviewing *Meet Me in St. Louis* on November 25, 1944, critic James Agee wrote, "This habit of sumptuous idealization seriously reduces the value even of the few scenes on which I chiefly base my liking for the picture; but at the same time, and for that matter nearly all the time, it gives you, for once, something most unusually pretty to watch."

A perfect day at the fair could be ruined in many ways, by rain, cold, heat, exhaustion, not to mention the usual human foibles and follies. The expense was not trivial; expectations had to be met—a lot was riding on the day. Countless diaries, photos, letters, articles, and books exist—an obsessive amount of documentation. Visitors struggled to render the fair, to capture it, to wrestle it into posterity, to fix it in amber. As the final days neared, visitors undertook frantic tours to try to see it all again. But, as one fairgoer put it, the fair resisted narrative—it "cannot be even hinted at by words." Across accounts, the one that seems to crop up most often: "indescribable."

And so beneath the fairgoer's wonder was a kind of manic sorrow, a present-tense nostalgia, the fleeting realization that one couldn't live forever at this pitch. Dazzled by the lights, you already began picturing life outside the glare. You knew you were being changed, not always for the better. As one concert attendee rued in his diary, "Our taste will be better than our opportunities hereafter."

I sense the lights dimming—I'm getting frantic. Like any good con on the Pike, I've focused your gaze to suit my own ends, but what about the attractions I have left out? Bellefontaine and Calvary Cemeteries, those radiant cities of the dead, the final resting place of Fair President Francis—their massive mausoleums impress even today, but I am growing tired of plans that look toward eternity. *(Francis ran his operations from an office across the quad from mine—the smell of champagne will last for months after he's gone!)* And what about other subterranean histories, such as the city's natural caverns, which—offering cool storage—built the city's beer business and later hid speakeasies, roller rinks, and theaters that offered light opera to audiences of three hundred? I haven't mentioned the local earthquakes—some of the greatest recorded on the continent. Or the indescribable City Museum, with its surreal six hundred thousand square feet. *Meanwhile, Frank Lloyd Wright is becoming enamored of Japanese design at the fair.* What about the city's Bosnian community flourishing on the south side, now the largest (per capita) outside Bosnia itself? *Or the Liberty Bell, delivered to the fair after seventy-five thousand local children signed a petition, and now guarded by a real Philadelphia policeman?* Whither the city's literary sons and daughters: Marianne Moore, William S. Burroughs, Jonathan Franzen, Kate Chopin *(who collapsed at the fair and died two days later),* and Tennessee Williams, who—while institutionalized in 1969 in a hospital on the east side of Forest Park—began assembling a secret time capsule, an olive-green Samsonite suitcase into which he put journals, poems, parking receipts, and his first passport, before leaving it with his personal secretary and lover. Layers upon layers! *At the Aeronautic Concourse, where the "bird-men" dock their floating vessels behind a thirty-foot fence, none of the airships will navigate the course fast enough to collect the $100,000 prize.* And what about the *Spirit of St. Louis,* Lindbergh's plane backed by businessmen from this fine town, which the former mail pilot predicted would become the country's aviation center—another promise unfulfilled! Or the amusements of the Pike, which might have become permanent had Washington University not protested the distraction: the Temple of Mirth, the Spectatorium, Creation, the Hereafter. *Thomas Jefferson's original headstone has been brought to the fair. Outside the Alaska Building, union* *workers are planting some newly painted totem poles upside down.* The city has known horrors—it, incidentally, is the birthplace of Vincent Price and the town where the man Scotland Yard believed to be the prime suspect for Jack the Ripper drew his last breath. *U.S. Marines will be drilling from 8:00 until 10:00 a.m., followed by feeding hours for the seals.* Oh, and I have said nothing of Nelly.

Meet Me in St. Louis ends at the fair. Overlooking the Grand Basin, Garland swoons to her beau, "I never dreamed anything could be so beautiful." When the palace lights come on, her mother sighs, "There's never been anything like it in the whole world." The youngest sister asks, "Grandpa? They'll never tear it down, will they?" He replies, "Well, they'd better not." Garland gets the last breathless word: "I can't believe it. Right here where we live. Right here in St. Louis." Fade out.

The fair closed on December 1, 1904. The governor declared Closing Day a school and business holiday. As midnight approached, President Francis made a brief speech, then turned off the lights as the band played "Auld Lang Syne." His face floated over the grounds, painted in fireworks next to the words "farewell" and "good night."

Having cost roughly $50 million to build, the disposable fair buildings were sold for $450,000 to the Chicago House Wrecking Company, which salvaged and resold a hundred million linear feet of lumber ("enough to build outright over ten cities with a population each of 5,000 inhabitants"), plus roofing, steel, doors, plumbing, fittings, and so on. Also for sale, some 350,000 incandescent lamps: six cents used, sixteen cents new. Several architectural forms appeared in Missouri homes.

In an Alpine-themed restaurant on the Pike, the fair fathers pick their teeth with quail bones; juice from medallions of beef drips down their chins. In the corner, the governors of Wisconsin, Missouri, and Minnesota share a hearty joke. In five days, President Francis will turn off the lights. The men are confident and rich. One of the waiters bumps me. He says to a guest, "Pardon him, sir—just another rube at the fair." The gentlemen smirk in my direction before offering me a seat. One asks, "Well, was it worth it?" Does he mean the fair or my visit? Then President Roosevelt stands and delivers these thoughts: "I have but one regret, and that is a deep regret—the regret that these buildings and these exhibits could not be made permanent; that these buildings cannot be maintained as they are for our children and our children's children and all who are to come after, as a permanent memorial of the greatness of this country. I think that an American who begrudges a dollar that has been spent here is not so far-sighted as he should be. It is a credit to the United States to have had such an exposition carried on so successfully from the beginning to its conclusion."

Many people mistake the Gateway Arch for a parabola, but it's more complicated than that. In truth, our arch is a flattened catenary arch—the graph of the hyperbolic cosine made flesh, or, put another way, the natural curve a cable makes when suspended between two points. The name comes from the Latin word for *chain.* I have ridden a tiny tram capsule up the north leg of the arch. The 1960s space-age pod barely fits five people and rises in a rocking-step motion—ka-chunk, ka-chunk—followed by stretches of smooth ascent. Travel is not for the faint of heart, nor the claustrophobic. My head brushes the roof of the pod. Before boarding, a futuristic female voice speaks from a monitor: "The Gateway Arch transcends time." The trip lasts four minutes.

A Park Ranger repeats over and over: "Welcome to the top. It's all good," and, "You got questions? Let your tax dollars sing." Visitors get their bearings by peering through seven-inch slits. A hot summer day: the arch casts a long, lopsided shadow. It takes a moment to orient yourself; approaching the arch through the Jefferson National Expansion Memorial park, I heard an excited girl shout, "To the west!" when we were actually migrating south. The base of the arch holds something called the Museum of Westward Expansion.

The arch is centered on the Old Courthouse, immediately to the west, where slaves were auctioned and Dred Scott tried his case twice. Ahead of me stretches downtown, Pruitt-Igoe, Forest Park, my university, my neighborhood, a number of private places. What was once the world's largest wheel no longer rises over the fairgrounds—after the fair, it was dynamited, its countless perfect spokes twisted and heaped like the slack tendrils of some broken beast. The remains were sold for scrap, but the wheel's seventy-ton axle—then the largest piece of forged steel in the world—remains missing to this day. My mind turns circles. I look over the north side and picture Dragon Chop Suey, just a few blocks from the corner where gunfire erupted last night as part of a citywide spree of unrelated violence. Seventeen people shot and one man stabbed—there have been no arrests. The police chief was out of town and had no comment for the paper.

Behind me stretches the swollen, brown Mississippi, well above flood stage, having already swallowed the lip of downtown. On a submerged street, a stop sign lists with the current. Muddy water laps up the steps leading to the arch, as if to reclaim it. Across the river rise a casino, grain elevators, an American flag, and the train yards and telephone poles of East St. Louis.

Most of Pruitt-Igoe has returned to the wilderness. The land sat unused until 1989, when fourteen acres became a public school site that still houses magnet middle and elementary schools. The remaining thirty-three acres have become an abandoned urban forest bound by an easily and—judging by the total collapse in some places—frequently scaled chain-link fence. An access road leads to an electrical substation on the site; a chain dangles between two short poles, blocking the way, another gate swung shut: DANGER: KEEP OUT.

Trash and debris collect beneath tall, fast-growing trees. A crumpled beer can—the label says Busch, a local name. Sirens are approaching, but no need for alarm—there's a fire station adjacent. Weeds and grass have taken over the cracked pavement that has become a wooded path; here and there a manhole pocks the forest floor. Further on, a storm drain disappears to nowhere. A streetlight sprouts from a lonely copse of trees. Former residents still stop outside the fence, where honeysuckle grows. They miss their radiant city. They dream of it. Part of them still calls it home.

Even after the fair ended, visitors returned to wander the empty paths. One woman wrote her husband that she was "heart sick to see the ruin and desolation"—but she could almost imagine herself still at the fair.

Here's the thing: the arch is beautiful. Before moving to town, I dismissed it as a hulking piece of modern midcentury kitsch—a civic branding tool, the stuff of bad airport T-shirts and mugs. But for more than a year the arch has watched over me, and while there are places in my neighborhood and on campus where I know to look for it, I often find myself catching an unexpected glimpse and—shocked back down to size—experience a jolt of the sublime. It is not unlike what used to happen with the Twin Towers. The arch is machined, perfect, soaring—the city's greatest open gate. It changes color with the weather and hour: sometimes sky blue, sometimes gunmetal gray, sometimes pitch black. Hovering on the horizon, it has its own moods. But the arch can be a sad star to steer by. I can't help but remember it's just an upside-down chain.

It's time to leave, but I linger. I know less than before, but as another fair-goer will write in his memoir, "Oh it was so sweet to sit there and look and listen, why could it not last forever? Who wanted to think of going home? Home was a fool."

On August 9, 2014—six months after a version of this essay is published—unarmed black teenager Michael Brown is fatally shot by white police officer Darren Wilson in Ferguson, Missouri, a small suburb in the northern county of St. Louis, just outside the city proper. The demographics of Ferguson have changed dramatically in the past twenty years, as black residents have resettled there from St. Louis and white residents have fled to further exurbs. Despite the shift in population, Ferguson's power structure remains white. The shooting outrages the communities of Ferguson and of St. Louis in general. Three months later, a grand jury decides not to indict Officer Wilson, and protests spread from St. Louis across the country. The signs read: BLACK LIVES MATTER.

The arch has no keystone; the north and south legs are of equal length. You're either on one side or the other. Arches, it should be noted, hold themselves up: they rise on their own weight, they compress—higher pieces push down and out on those below. Some five hundred tons of pressure were needed to pry the legs apart to install the final four-foot piece. That's why the windows are so small: to preserve the structural integrity. A wider view would cause the whole thing to crash down. The National Park website lists the exact mathematical equation that describes the arch, but it never was a pure construct: it sits eighteen degrees askew from the north–south axis and sways some eighteen inches, like a chain or a gate. That fact does not comfort me when the wind blows, or even on a clear day like today. The balance is an illusion. I stand at a tense threshold—atop the tallest man-made monument in the country—upheld, for now, by forces great and unseen.

2014

End of the Line

1

One Sunday in May 2008, after signing a dubious waiver, I followed my wife and two friends down a ladder into a manhole in the middle of downtown Brooklyn's Atlantic Avenue. Some orange cones and two striped sawhorses erected by a large, taciturn tattooed man diverted traffic around us. A small portable generator snaked a power cord into the hole. Onlookers gathered on the corner by the bank.

We dropped into a short trench dug out of sandy rock. The air was close, the space claustrophobic. Through a hole in the far wall, we descended a long makeshift stairway that led to the tunnel floor. The tunnel was twenty-one feet wide and seventeen feet high—a vast chamber that stretched into the darkness. Thick, blue stone walls gave way to a stunning brick arch that barreled above us. It was warm, humid, and deathly quiet. We were stories beneath the busy street.

Built in 1844, this half-mile passage can be considered the world's first subway tunnel. Steam trains—two abreast—shuttled passengers through downtown Brooklyn between South Ferry landing and a station of the Long Island Railroad. The Atlantic Avenue tunnel was sealed in 1861, after steam trains were banned in Brooklyn. But the dishonest contractor—paid to fill in the tunnel—only capped the ends, one of which I had just stepped through.

I carried a flashlight. My friend wore a headlamp. We could see patches of the original whitewash on the vault. A chain ladder hung on a wall. A wheelbarrow held trash and pieces of wood. We stepped around blocks of masonry, the remnants of street-level ventilation shafts that, during demolition, had collapsed to the tunnel floor. On one block sat an old black telephone with a red label: SPEAK LOUD. Dark specks dotted the bricks on the ceiling—150-year-old soot, our guide told us. The rails were gone, but the dirt floor held telltale mounds and hollows, faint footprints where the ties once ran.

The tunnel had been lost and found many times. A search party organized by a newspaper couldn't find it in 1911, but the wall bore graffiti from workers who broke through the roof in 1916—either G-men chasing rumors of subterranean German saboteurs or a telephone crew laying cable, depending on whom you asked. Two decades later, the police—tipped off by an anonymous note—went looking for a gangster's body thought to be stashed in the tunnel, but they never found a way in. Bob Diamond, a train enthusiast, rediscovered the tunnel in 1980 when he went down a blank manhole cover marked by a blue dot on an old map. He eventually dug his way seventy feet to a wall and broke through to the main chamber.

Bob was our guide. Wearing jeans, a black T-shirt, and work gloves, he led our group into the dark. He told stories as we walked beneath some three city blocks between Court and Hicks Streets before hitting a bricked-up bulkhead, behind which he believed might be buried an antique locomotive. Lights swept across some debris. A camera flash illuminated the vault. Constellations of mold clung to the ceiling; spindly multiheaded fungi sprouted from cracks in the walls. In 1844, while the tunnel was being built, a cave-in killed a man named Collins and his horse; an Irish laborer shot his overseer; and a pedestrian fell into tunnel construction and died. Two years later, police were forced to investigate complaints of an underground ghost. In 1893, decades after the passage was closed, the *New York Times* ran a fictional story about a gang of river pirates hiding loot in tunnel. Over the years it has been said to house bootleggers, mushroom caves, counterfeiters, overgrown rats, and—in a 1925 short story by H. P. Lovecraft, who lived a block away—demons and devil worshippers.

Bob sat on a stone block as we clambered up the dirt piled against the far wall. He was desperate to break through or tunnel behind it. Four years later, he would learn engineering consultants using electromagnetic sensors had remotely identified a twenty-foot-long "subsurface metallic anomaly" beneath the street at that end of the tunnel. At last, his locomotive! But by then the Department of Transportation—citing safety concerns—had ended his tours and closed the tunnel once again. Agents of the city would weld his manhole shut.

But that day, we surfaced, blinking, happy, enlivened by mystery, and went on with our lives. Looking at the photographs I took, I'm caught a little short. My wife and I don't live in Brooklyn anymore; we left three months after that underground trip. My friend and his girlfriend—who mug for my camera—moved away, moved back, moved away again. He got sick; they broke up. He went to Texas. Now my wife and I live in St. Louis. To this day, Bob cannot go back into his tunnel. We cannot go back either.

2

Summer of 2000: I was twenty-three years old when I moved to New York, to a rundown residential hotel on the Upper West Side. In those early days, everything seemed "underground"—hidden, secret. There were underground economies—*why did the cashier at the bodega wink when he slipped me a free bagel*—underground parties—*was the cute girl at work* (whom I later would marry) *kidding when she said she went to an*

SNL cast party at a club in a subway station?—underground art—*was it lame to be moved by* Macbeth *performed by inch-high plastic ninjas?*—underground etiquette—*do you acknowledge it when a stranger sneezes on the train?*—and even more underground economies. I remember a would-be actor friend who took an apartment that—in exchange for cheap rent—required him not only to feed but to spend a certain amount of time playing with his landlady's cat. As my friend crouched on his knees, waving his hands at the disinterested feline, the woman told him, "Feeding Catmandu is only half the job; the other half is loving him." At the time, it seemed like a fair bargain.

I was even luckier than my friend: within a few weeks of moving to town, I got a job at a magazine. My first assignment was to interview a fading socialite. I took a cab to the Upper East Side, giddy with the idea of being able to expense the ride, double-checking the batteries in the tape recorder that I had just spent more money on than practically anything else in my underfurnished apartment. I remember sitting on the family's impeccably tasteful and (as I would learn in the coming years) perfectly predictable period French furniture, the white molding in the ceiling as straight as the piping on the doorman's jacket, and listening to stories about the Kennedys and dazzling island parties where everyone always ended up in the ocean. The dog was named after an old screen star—I don't recall which—and the bookshelves held faded copies of *Life* magazine. Here I learned a new phrase—"the royals"—that rang as startling music to a boy like me from Texas. It was the socialite's grown daughter who said it—she who had been no more than knee-high during the family's golden age—and she said it with such sadness: "I missed all the celebrities—I was born afterward. But I remember the royals. . . ." I can still hear her voice trailing off. I had never before heard an ellipsis spoken aloud.

Not long after the closing of the Atlantic Avenue tunnel, Walt Whitman mourned its loss in the *Brooklyn Daily Standard:*

> The old tunnel, that used to lie there under ground, a passage of Acheron-like solemnity and darkness, now all closed and filled up, and soon to be utterly forgotten, with all its reminiscences; of which, however, there will, for a few years yet be many dear ones, to not a few Brooklynites, New Yorkers, and promiscuous crowds besides. . . . The tunnel: dark as the grave, cold, damp, and silent. How beautiful look Earth and Heaven again, as we emerge from the gloom! It might not be unprofitable, now and then, to send us mortals—the dissatisfied ones, at least, and that's a large proportion—into some tunnel of several days' journey. We'd perhaps grumble less afterward at God's handiwork.

But Whitman's silent solemnity doesn't quite match a report from the *Brooklyn Eagle,* which, upon the tunnel's opening, described the inaugural trip taken by railroad and political dignitaries:

> The sonorous puffs of the engines; the clattering and echoes of the cars, reverberating through the cavern; and the deafening and uproarious shouts of the company—it might safely be characterized as an underground swell. . . . The darkness and smoke were so intense and pervading that no one but an emigrant from "Pluto's dark domain" could have seen a foot beyond his nose.

Thus a choked, tumultuous journey is smoothed out in the cold light of memory.

Johannes Hofer, the Swiss doctor who first named *nostalgia* in 1688, called it a "cerebral disease of essentially demonic cause"—the "continuous vibration of animal spirits through those fibers of the middle brain in which impressed traces of ideas of the Fatherland still cling."

Nostalgia was historically understood as a kind of physically debilitating homesickness—sufferers simply wasted away. The cure: bring the patient home. (Not surprisingly, this remedy was attractive to soldiers fighting abroad, many of whom, when diagnosed, were allowed to return from the front.) While Hofer studied the symptoms in Swiss mercenaries, it was a Russian army officer who, in 1733, is said to have curtailed a nostalgic outbreak among his troops by burying one suffering soldier alive.

But recent psychological studies have pointed out nostalgia's positive effects: the afflicted feel less lonely and bored, more happy and generous. Chinese researchers have found that cold temperatures can bring on nostalgia, which makes you feel physically warmer. Nostalgia thaws something inside you. Of course nostalgia is predicated on a large amount of forgetting.

Walter Benjamin wrote, "The delight of the city-dweller is not so much love at first sight as love at last sight." What does it mean to say goodbye to New York? That question has spawned its own genre, the Leaving New York Essay, which follows its own formula: (1) evoke a bygone, breathless "scene" that cannot be reclaimed but in memory (downtown, the East Village, the Lower East Side, pre-gentrified Brooklyn); (2) express a frustrated desire to stop time and stake your place; (3) indulge in nostalgia, however covert, (4) insert disillusionment, (5) distance yourself from New York (emotionally, if not physically). Deep down, these essays—and there seem to be anthologies published every year—are not really about a specific city but about lost and romanticized youth. *Blink—I grew up!* Of course, this piece is my own contribution to the genre, for it was at that first magazine job that I gave my wife-to-be Joan Didion's archetypal Leaving New York Essay "Goodbye to All That," which we read along with *Variety*—like young Didion—while working late and waiting for the copy desk to call. Also like Didion, I would stay at the fair for eight years.

Is it just nostalgia, then? The narrative can be so simple—from happiness to despair to (if you're lucky) a different kind of happiness—though for me the point of "Goodbye to All That," which my writing students sometimes miss, is that Didion is romanticizing her flight to dreamy Los Angeles in the same way she first idealized New York City. She has traded one romance for another, and my students are crushed when I tell them she ended up moving back to New York.

Innocence lost, a less dazzling adulthood gained—the lament of the sad old man at the bar, the stuff of clichés. The editor who assigned me that first magazine piece was brilliant and exacting, an arbiter of taste, architecture, and design with a disdain for the false, indulgent, and sentimental. Blunt and opinionated, he commanded—some might say terrorized—the big glossy magazines. A fiercely private man, but a real city creature, his choice in office chairs would appear in the gossip column of *New York* magazine. I slowly got to know him, and we worked together on two magazines that both folded. He would assign a piece saying, "Be poetic and brilliant" and then—as the impossibly short deadline neared—send goading emails: "I'm facing a blank screen"; minutes later, "No rush, really. Just any time." He would then edit the piece into, if not poetry and brilliance, at least better prose than the subject—a recent runway collection, a hidden Tasmanian resort, an appreciation of the espadrille, some fresh Hollywood face, the latest passing fancy—might deserve.

Then, improbably, impossibly, he sold his elegant and airy prewar Greenwich Village apartment and left New York: moved west, where he volunteered his time tutoring high school English (none of us could imagine this—he who edited so ruthlessly!), then moved back east, settling in Boston, in what must have been for him a dullish self-exile. When my wife—who also worked with him—gave birth to our daughter, a package arrived on our doorstep in Minneapolis, site of our own New York exile. It contained a carefully wrapped set of antique girls' clothes—the white linen faded to parchment—a curiously sweet outfit. The note read, "I found these in a little French antiques shop, and, though absurd, I thought your daughter might not be so jaded yet." Maybe sentiment finds us all.

The next correspondence came six weeks later in the form of an email from a lawyer. Two lines, without salutation or sign-off, stating that my friend had died the night before. There would be no funeral or memorial service. There was also no obituary, no notice in the newspapers and magazines that had once tracked his moves and whims with such glee. To put it in phrasing he surely would have hated, he disappeared without warning, without a trace.

Afterward, a New York editor and I emailed about how we wished someone would write about our mutual friend and his many contributions—we both had learned so much from him—but we were crippled by the knowledge that he would have hated to have people gossiping about him once again.

In *The Colossus of New York,* Colson Whitehead writes, "You are a New Yorker when what was there before is more real and solid than what is here now." To be a city dweller is to conduct a love affair with a past that is being endlessly eroded, to inhabit a string of vanishing cities stretching down to the bedrock, to be a captive of loss. On the flip side, Whitehead warns, perhaps more sensibly, "New York does not hold our former selves against us. Perhaps we can extend it the same courtesy."

Three more stories about the slipperiness of memory:

In the weeks leading up to my daughter's third birthday, she became gripped by nostalgia for her own infancy. She began insisting she was a baby—she needed a crib, not the big-girl bed she had begged us for; she made us carry her around while she cooed in a made-up baby babble; she sighed and asked pensive questions about the particulars of her past ("Did I sleep with *this blanket* when I was a baby?").

In April 2001, I cut out an article from the *New Yorker* about the intended demolition of El Teddy's, the landmark downtown restaurant.

I underlined this sentence by the writer, Adam Gopnik: "New York has always been a place where it is possible to have memories without the experiences that conventionally precede them." The year before, when tasked to think up a date night for my office crush/future wife, I told her I thought we might begin—as countless young romantics before us—by meeting under the clock in the lobby of the Biltmore Hotel, which, I was subsequently sorry to learn, had been gutted two decades earlier and turned into a bank. It was early in the relationship. We went tango dancing instead.

Researchers have proven that mice can pass down memories—in one experiment, an aversion to the smell of cherry blossoms—through at least two generations. The mice's children, and their children, inherited the fear—a genetic memory—despite never having encountered the smell. Trauma slips through the ether. Going further, we can even create false memories in mice by activating specific parts of the brain, making them later "remember" that they had been shocked in one location when it was really elsewhere. The process turns out to be relatively easy. One article noted that, in terms of the cognitive mechanics in play, we are quite similar to mice. That the brain is wired to deceive us—to conflate past and present—seems to be an evolutionary advantage.

3

New York is a city that connects best underground. Five railroads tunnel beneath it (Amtrak, Metro-North, New Jersey Transit, the PATH, the LIRR), navigating the subterranean web that undergirds the city: steam tunnels, power lines, water mains, storm drains, sewers, cable/data lines, and New York's staggeringly beautiful century-old system of water tunnels that—descending more than 125 miles from the Catskills and beyond—approach the city, in places, more than a thousand feet below sea level. (The aqueducts drop an average of thirteen inches per mile; in the city, water is delivered mainly by gravity, and the pressure must be reduced before surfacing—else there is risk of a blowout. Even then, the force remains sufficient to bring water to the sixth floor of most buildings.)

And then there's the subway. Running twenty-four hours a day, seven days a week, the New York City subway system boasts 6,311 cars running on twenty-four lines through 468 stations along 659 miles of passenger track that—if uncoiled—would stretch past Detroit. (Add the rails in the yards, storage, and maintenance areas and you'd get beyond Chicago.) More than five million people ride the subway on an average weekday. (The 4-5-6 line alone carries more riders daily than the Boston, San Francisco, and Chicago subways put together.) Fourteen underwater tunnels cross the city's rivers. Pumps remove thirteen million gallons of water a day from the system. And the subway continues to grow, burrowing a new East Side line beneath Second Avenue and extending the 7 line a mile and a half west of midtown to a terminus 150 feet belowground, for which a cave wider than the Empire State Building is tall was carved out of bedrock.

Such projects are a gold mine for geologists eager to plumb the depths of New York. Much of the city is built upon the Manhattan schist that anchors midtown and downtown, rock folded and formed by heat and pressure born from the collision of continents. Tony Hiss, author and urbanist, described the excavation to the *Times:* "New York's deepest and darkest secret, its oldest and most violent and previously only vaguely glimpsed history is finally coming to light."

In 1854, a man "about 5 feet 6 or 7 inches" wearing "greyish pantaloons" was run over by a train in the Atlantic Avenue tunnel. Witnesses saw him leap from the first car and run beside the train before slipping under the wheels. A daredevil? A suicide? The train was coming from the racetrack. According to the *Brooklyn Eagle:*

> The unfortunate man was cut in two, the upper portion of the head being completely severed. Assistant Captain

> Brown, and several officers of the First District Police, who had come in the same train, took up the body, and had it conveyed to the Dead House, in a cart.

I was once on a subway train that ran over a jumper. I was in the middle car and happened to be talking to the conductor when the train screeched to a halt. The intercom crackled. "Fuck, fuck, shit," the conductor said. "Twelve-nine, person hit by train." I must have looked shocked, because he added, "I see this four to five times a week." I didn't think I believed him. We were between Delancey and Second Avenue. Most of the train had stopped short of the station, and as we were evacuated out of the tunnel and onto the platform, I heard a witness tell the police, "She flung out her arms." Flashlights played over the tracks. At the edge of the platform, a little girl stood, stone-faced. She must have seen everything. They gave us free transfers, but I went to the surface and got into a cab that was playing "Here Comes the Sun."

I also remember the collective denial of standing in a late-night crowd on a downtown platform and watching a homeless man jump down to pick a quarter off the tracks.

Then again, I can recall returning from a trip down south and feeling sad to be back in cold and rainy New York until I sat down on a train next to a tall fortysomething man in a dark, expensive-looking suit who happened to be wearing an enormous pair of pink, fluffy bunny ears.

I remember having a free birthday party at a bowling alley because a friend had gotten beat up in the ladies' room the week before and didn't press charges. On weekends, we'd have to move the pickup street-soccer game in Central Park further down the asphalt so the drug dealers could have their usual spot next to the roller-disco party. ("You gonna keep me from feeding my family?") After a nasty tackle, one player broke a beer bottle on the curb and went after his opponent with the jagged neck—the guys on the sidelines separated them while we kicked glass off the "field." Sometimes I can still smell the sweet stench of Icelandic poppies rotting in a two-foot brass Indonesian vase sent to me at the magazine from a desperate publicist. I remember my wife's old roommate riding all the stops of the subway on a single fare—the entire route taking him thirty hours and thirty-three minutes. Meanwhile, I tried out for a game show again and again, which seemed like smart financial planning. I remember spending a night acting as a "traffic light" at a dance rehearsal, which meant it was my job to say "green," wait sixty seconds, say "yellow," wait sixty seconds, say "green," wait sixty seconds, and so on for three hours, there inexplicably being no "red"—all for eighty dollars. I remember holding the door for Kurt Vonnegut, who was wandering out of a screening as I walked in; a few minutes later a woman rushed up and gasped, "Did Kurt just *leave*?" Celebrities dropped by our office; I once left the toilet seat up on Salma Hayek, who went after me into the unisex bathroom. I remember standing forever on First Avenue at 4:00 a.m., waiting for an off-duty cab to take me uptown on the way to dispatch. I did that a lot. I remember going to my first piano bar (Rose's Turn, now gone). I remember New Year's Eves spent at parties, in bars, walking the streets. One Fourth of July, I climbed a rooftop in Queens to watch fireworks, but that year they were in a slightly different location—and just out of sight. (The colors lit up the undersides of clouds.) At work, I was surprised to learn the chimp in a photo shoot made $800 a day, which meant in about a month it could earn the yearly salary of us young editors and perhaps even the finance assistant, who told me she didn't know whether to itemize the chimp as a model, prop, or equipment. One day, I set my VCR to tape the start of the war before going out for pizza. I remember the anonymous joker who would stencil a silhouette of a young Bill Murray on the ubiquitous blue construction fences that were always stamped "Post No Bills." After a long night with some friends, I remember telling a bartender in an East Village dive that 4:45 a.m. was no time to be closing a bar in New York City. The guy didn't look up but just said, "Come on, it's Sunday." Another night, I boiled lobsters on a hot plate in my kitchen, which had been converted from a hall closet. I remember early days riding the subway, opening my jaw slightly to feel the hum in my teeth.

Ian Frazier once wrote in the *New Yorker*, "The urge to tunnel is partly an urge to disappear, and its product, no matter how monumental, is visible only from the inside." But what about things buried, that exist under pressure—don't they want to erupt into the light?

I remember reading E. B. White before the planes hit: "All dwellers in cities must live with the stubborn fact of annihilation." In those days, I got dressed in the mornings with the TV tuned to NY1. That Tuesday, there was a fire in one of the Trade Center towers; witnesses said a plane had hit it; strange for such a clear day. Then the second tower exploded. My mom called, crying. She asked where my girlfriend was. I said she was in her East Village apartment. From the other room, my roommate (watching a different angle on another channel) said he saw it—definitely a second plane, not just an "explosion," as the newscaster claimed. I hung up and left a message on my girlfriend's answering machine, telling her to turn on the TV, on which a voice was wondering whether the second tower might have been hit by a news or police copter. I caught a downtown C train since the 1/9 was closed, the idea being to get to the magazine's office in Chelsea as fast as I could. There we would make sense of the news.

4

At four o'clock on the afternoon of February 19, 1916, twenty-eight-year-old sandhog Marshall Mabey was working below the East River on a subway tunnel to connect Montague Street in Brooklyn Heights to Whitehall Street at Manhattan's tip. River tubes were being cut using the "shield method," in which a sharp steel plate slowly carved through the bed of the river. Sandhogs removed earth at the head of the shield, which was advanced by hydraulic jacks twenty-six inches at a time—at which point another steel ring would be added to the tube growing behind it. Ahead of the rings, wooden planks propped up the loose dirt, but the real work of keeping the soft river bottom from collapsing down on the miners was done by the compressed air pumped into the tunnel—on that day, twenty-four pounds per square inch, or three times the pressure of an NBA basketball. Under such compression, miners would get the bends, and it was impossible to whistle.

Mabey was some six to seven hundred feet from the Brooklyn shore. The shield was ready to be forced forward, and he and three others were removing the boards shoring up the front of the tunnel, when a rock twice the size of a baseball fell out of the ceiling. Mabey's partner tried to plug the widening hole with a sandbag, but—with a sudden *crack!*—the man disappeared. Mabey lunged for the pipes pumping air into the tunnel, but couldn't grab hold as he, too, was sucked toward the roaring breach in the ceiling, which he estimated to be about eighteen inches. He later would tell the *Times,* "As I struck the mud it felt as if something was squeezing me tighter than I had ever been squeezed." He blacked out as he was shot through twelve feet of riverbed and then through the East River before being launched twenty-five feet into the air.

A front had blown in—the thermometer read eighteen degrees, cold even for February. Witnesses along the waterfront and the Brooklyn Bridge were shocked to see three men suddenly erupt from a two-story geyser before splashing back into the river. Mabey came to. "I am a good swimmer," he reported, "and I kept my mouth shut and came up to the surface." His rubber boots weighed him down and his left leg felt numb (it was later found to be broken), but Mabey swam to a rope thrown by men on a tugboat. The two other sandhogs shot through the hole were not as lucky. Fifteen minutes later, someone spotted Mabey's partner, who was dragged by a boat hook onto a sand scow. He died soon after, having had struck his head on the bottom of a barge. The next day, harbor police found the body of the third man, who had drowned.

It was the longest, quietest subway ride of my life. The car was full but not packed. I didn't have a seat, but I leaned against one of the poles after about an hour. Stopped in the tunnel, there was nothing to see; the windows were dark. Nobody spoke or made eye contact as the voice repeated again and again that we were being delayed "due to a police investigation at Chambers Street." I was grateful for the silence, because it seemed that if anyone spoke—that is, acknowledged whatever it was that was going on—panic would break out. Eventually we were evacuated into Grand Central Station, though I have only my notes from that day to go by, as I don't remember this at all. Penn Station seems far more likely. I remember surfacing, the streets overflowing with people walking against traffic and lights, stopping to huddle around cars stalled in their spots, radios blaring, windows down for everyone to hear. Crowds were plastered to the windows of electronic stores. Lines stretched from pay phones. People were crying and carrying gallons of water. Sirens. Smoke. An indescribable smell, one that would linger for months. My cell phone briefly got service; from my mom I learned the towers had fallen and the Pentagon had been hit. At the sound of a plane, all civilians looked up.

I walked south to the office, where you could get a phone call out every third try. Someone on the business side of the magazine said, "It's going to be a motherfucker for the economy." He claimed to have called his broker the minute the first plane hit and exhorted everyone within earshot—mostly young assistants trying to call their families—to do the same. We had run a story on bin Laden a year earlier, and one of the editors was frantically trying to dial the U.K. to get in touch with the writer of a follow-up piece that recently had been scrapped. Most people ignored her and gathered around the TV in the conference room. I heard my boss's line ring, and I ran and got her—she cried at the sound of her boyfriend's voice. I had a message from my girlfriend, who had walked to work and then taken a number of displaced coworkers back to her apartment. Everyone was going home. The whole city, it seemed, was out walking, steering itself by a pillar of smoke hung in the sky. On my way to the East Village, I watched two kids pose with it in a picture.

It goes without saying that my 9/11 was overwhelmingly benign compared to so many others' that day, but I want to say it anyway. There were those who died or lost loved ones. There were friends who had to pick their way through glass and fallen bodies. There was the former college suitemate who slept through his alarm and didn't show up to work on the top-floor offices of Cantor Fitzgerald. I remember crowding onto the couch in my girlfriend's tiny living room, listening to the TV doctor say again and again he didn't think chemical or biological weapons had been involved—"as far as we know"—and glancing at her roommate's bag, still covered in white dust on the floor. No one said anything or looked at the roommate. He'd come home hours earlier, after ducking under a car when the debris cloud rolled through the canyons of Wall Street. ("I couldn't see my hand in front of my face.") After a shower, the guy seemed fine, but he got drunk at dinner—at a barbecue place open around the corner—and drunker at home over beers with another roommate and his friends, who were already back from being turned away at the blood bank. As it grew dark, makeshift monuments sprung up, candles flickering on empty doorsteps. We slept with the windows open. It was quiet except for emergency vehicles, which were in a way comforting. The city was closed to traffic below Fourteenth Street. I had never noticed how wide the streets were without the parked cars.

Mabey returned to his Long Island City home a day after the accident, saying he hoped to be back to work in a day or two. His wife told the paper, "His job is a good one and I'm glad he has it." The tube remained partially flooded, though the air pumps stayed on and water bubbled four feet above the surface at the point of the break. A blanket of hundreds of tons of clay would be dumped from above to reinforce the riverbed. Natural weaknesses in the bed were to blame, according to the *Times*, which stated, "As yet engineers have found no way to discover these flaws until they reveal themselves."

Mabey spent the next twenty-five years on the job. Two of his five children would become sandhogs, too. The Montague Street Tunnel opened for business on August 1, 1920, and continues to be in use (by the R train). It flooded again during Hurricane Sandy, but reopened in September of this year.

In the following days and weeks, we worked a lot. The staff spent most of its time at the magazine. One day we made vats of pasta and sauce in the company kitchen to deliver to the searchers. We ripped up the issue that had just been sent to the printer. Photographs came in from the rubble. We were trying to turn a photo spread that had been about the "power tribes" of New York into a portrait of human resilience. Of course this made no sense—seemed blasphemous, really—but we pursued it with such zeal. It was easier than thinking. *Keep moving.* We interviewed socialites, financiers, photographers, artists, chefs, filmmakers, schoolteachers, firefighters, writers, doctors; we collected stories, stories, stories: *Where were you?* Someone got Chelsea Clinton to write an essay. But my interviews were strained, my notes nonsensical. I remember cold-calling the family of a beloved priest who—after administering to victims and firefighters—rushed into one of the towers and was crushed by debris. My hands shook as I dialed; someone picked up before I knew what to say. Earlier, I had interviewed the chief medical officer of St. Vincent's, the closest hospital, about how his staff was 100 percent mobilized and ready for patients—that is, survivors—that we both knew would never show up. I remember before this happened it had been summer, and now it was fall.

5

The inventor of one of the earliest tunneling shields was Alfred Ely Beach, editor of *Scientific American.* After his proposal to build an underground railway was denied, Beach applied for and was granted permission to create a subterranean pneumatic mail service in downtown Manhattan. Cleared to bore two four-and-a-half-foot mail tubes that would connect to the main post office, Beach petitioned to excavate a larger passage ostensibly meant to house both tubes—his eight-foot subway tunnel, built partly in secret, thanks to his newly devised shield, which did not disturb the surface. (The earth was carted away at night.) A classic bait and switch—Beach dug the subway he had been denied.

The pneumatic railway was unveiled on February 26, 1870. From the basement of Devlin's Clothing Store—at 260 Broadway, across from City Hall—Beach had built a 294-foot iron and white-brick tunnel that began beneath Warren Street and curved twenty-one feet below Broadway to Murray Street, a block away. A fifty-ton mechanical blower dubbed "the Western Tornado" forced the eighteen-passenger cylindrical car to the end of the tunnel—then sucked it right back.

I can almost recall the feverish paranoia of that time, a dizzy, slow-burning haze. Days of strange men handing out envelopes on the subway, of contagion in the mail, of walking past contaminated office buildings, of discussing how you might seal up your apartment windows with trash bags, if needed. We heard in the office just before it hit the news—Tom Brokaw's assistant had anthrax. The media was being targeted. (We were co-owned by a company owned by Disney, which also owned ABC, which had received a poisoned letter.) I was sent a stash of Cipro and iodine pills (in case of a dirty bomb). At that point our office mail was being irradiated *and* steamed; what letters survived the process arrived hopelessly late and nearly illegible, the clear plastic return address stickers in the upper corners shrunken into hard fossils, like the Shrinky Dinks we used to color and put in the oven as kids. Assistants opened mail for their bosses, who wondered if they should invest in gloves and masks. (The more humane editors began handling their own correspondence.) There were rumors, incidents, hoaxes, and scares. Some days, taking the train felt like an act of defiance, while duct tape bravely sold out in the suburbs. We'd check the TV for street and subway closures before heading home from work. I sent my parents short emails that were meant to relieve them. Two days after the attack: "Getting to work was a pain—again. They closed my subway line south of 42nd Street because they're worried about structural damage. I heard there was another bomb scare in another building—they evacuated Grand Central—and this morning they closed Staten Island. (The bridges were closed briefly again early today.) It's still very crazy. But it was nice to get home and change clothes, etc."

Then again, I remember the midwestern cheeriness of the guy from the credit-card company who—on a routine call—broke script and asked how things were in New York. He had been there once and had paid a guy thirty dollars to take him from JFK to LaGuardia and said I should come to Missouri now that turkey season was ending and deer season about to begin. The John Denver look-alike in Central Park had a bigger audience than ever that first Sunday after; I watched the people on the lawn close their eyes and sway together to "Peace Train."

The "atmospheric railroad" literally went nowhere—there was only one station—but it was a hit. Visitors flocked to pay twenty-five cents a ride, which—because the Beach Pneumatic Transit Company wasn't officially chartered to ferry passengers—was donated to the Union Home for the Orphans of Soldiers and Sailors. Riders waited in an opulent saloon outfitted with Greek statues, damask curtains, a chandelier, a grand piano, and a fountain stocked with goldfish. The entrance to the tunnel was flanked by bronze statues and crowned by red, white, and blue gaslights. The car was luxurious—plush seats, oval windows, bright zircon lamps—and, traveling on wheels, topped out at about ten miles an hour. When it wasn't running, visitors walked the length of the tunnel, where Broadway traffic could be heard trundling overhead.

Legislation was vetoed to extend the pneumatic line to Central Park, the financial crisis of 1873 delayed the company's effort to raise capital, and by the time Beach secured a franchise for a private passenger line, the elevated railroad had taken hold. The tunnel was sealed. The Devlin's building would burn to the ground. Beach died in 1896, eight years before the Interborough Rapid Transit Company opened Manhattan's first subway line, which ran from City Hall to West 145th Street.

A few weeks later, eight of us headed to Vermont to a friend's cabin. The leaves were not at their peak, but I was shocked by their color. The house had no electricity, water, or heat. Just a place to stay on a thin strip of land. We hiked a mountain and read by a lake. How calm we were; escaping the city felt like breathing again after not realizing you'd been holding your breath. There is a picture of everyone on the porch, all goofy smiles.

The windows were boarded up, the bedroom pitch-black and cold. It must have been after midnight when the screaming began. It sounded faint, far off. In my half sleep, I burrowed deeper under the covers to escape the noise—was it squeaking bedsprings, the sounds of love? Fighting? Crying? It went on all night. In the morning, my memory foggy, I lifted the shade on the darkened window, which had been boarded up from the outside. I shouted: squeezed between the pane and the planks was a vibrating throng of bats—hundreds, it seemed—eyes shut, bodies twitching, a thick jostling mosaic of fur and ears and wings huddled together blindly in the dark. They had no idea I was watching. It seemed they could never wake up.

Six weeks later another plane crashed, this time into a neighborhood in Queens—at work, we shared the uncomfortable déjà vu of gathering around the conference-room TV. Six months later, I was still having dreams—airships descending on the city, chunks of the island crumbling into the water, bombs in the subway that always went off.

Some have claimed Beach's curiosity to be the city's original subway, but we know better. When the pneumatic train debuted, Brooklyn's Atlantic Avenue tunnel had already been sealed for nine years. In 1912—eight years after the arrival of the IRT and forty-two years after Beach unveiled his marvel—the subway finally reached beneath lower Broadway. Contractors found Beach's tunnel intact, along with the remains of an original car. They took souvenirs, and, at last, the tunnel was destroyed. Mail, however, would be delivered pneumatically through Manhattan—just not in Beach's tubes—in canisters traveling thirty miles per hour through twenty-seven miles of pipe four to six feet beneath the street until 1953.

In time, fear would give way to anger. We would be lied to. We would go to war. I was furious at our leaders, who seemed intent on squandering whatever international goodwill had arisen out of the horror. I became irrationally proud to have written the intro to a silly photo spread that got our magazine banned from the Bush White House.

Already something of a worrier, I became more imaginative in my fears, an expert at anticipating worst-case scenarios. My wife bore this patiently even after we left New York. Years later, I would keep a sleeping bag, a flashlight, and a granola bar in the trunk of the car. A spring storm would have me heading to the basement about five times in the night to check for flooding, my only comfort coming from the realization that—should the giant silver maple in the front yard blow down on our house—my wife and I would be crushed, but the baby's room would be spared.

6

The subway holds so many ghosts: abandoned stations at West Ninety-First Street, East Eighteenth Street, Worth Street, and Myrtle Avenue; the gilded City Hall station empty but on view as the downtown 6 train turns around at the end of the line (just stay in your seat); the sealed lower levels of Bergen Street, Nevins Street, and Ninth Avenue, where graffiti blooms in the dark; and the Cortlandt Street station buried beneath the towers.

I don't usually take the C train when I'm back in New York—recently the Straphangers Campaign rated it the city's worst line (dirtiest, most breakdowns, longest wait between trains)—but I found myself on a C the last time I was there. Weekend service was disrupted; the train was running off schedule on another track. The car was surprisingly empty for a Sunday in SoHo—only four other people. An ad next to the window read GET OUTTA TOWN. Brooklyn-bound, we had to get off unexpectedly at Second Avenue, the station closest to my wife's old East Village apartment. She was sitting next to me; I imagine we were both surprised to find ourselves there once again. Over the speaker, a woman's voice: "Last stop, last stop." The brakes let out a loud, final hiss. Passengers looked up, but no one moved. Then the conductor shouted, "This train isn't going *anywhere.*" So we crossed the platform, got onto one of the newer F trains, and went on our way.

Earlier that morning, I visited the Atlantic Avenue manhole that we had climbed down six years before. Another spring Sunday; the trees bloomed bright white. Back then they had been green and leafy, having already lost their buds. Now there was a Barney's down the block, and on the corner where we had lined up stretched a ramp to a no-longer-new Trader Joe's. Standing in the middle of the street, I could see the still-shiny welds that held down the dull cast-iron cover. The crosswalk beeped at me. I thought of Bob. I thought of old friends.

Traffic slowly turned onto Court Street as an Italian grandmother banged her shopping cart up over the curb, not heading to Trader Joe's. The day before, I passed my old landlord on the street; we didn't acknowledge each other. Today there was a guy selling records out of crates and some shops I didn't recognize: wine, organic burgers. A run-down Mexican place I'd been meaning to try for about a decade had gone out of business. A construction permit hung in the window. As she approached the grocery store, a young woman—hip, dressed for brunch—stopped at the corner and chirped to her companion, "At last, here we are!"

The September 11 Memorial Museum at Ground Zero finally opened this year. Nearly all of it is buried beneath the footprint of the towers, embedded in the Manhattan schist seventy feet belowground. Sections of the original foundations have been excavated. The concrete slurry wall—that bordered the site and miraculously kept the subway system from flooding that day—is exposed in "Foundation Hall," where it still holds back the Hudson. In an alcove below the ghost of the north tower stands a heavy fragment of impossibly dense strata—the contents, walls, and furnishings of many floors compressed and fused into a few terrible feet. The museum estimated 2.5 million people would visit in its first year, myself not among them. I am unnerved by so much history pressing overhead.

Most likely the longest I'll ever spend belowground will be that September 11th. I think of that ride often. The mute black tunnel. The fluorescent car. I wonder about my fellow passengers, where they are, what they've moved on to. I still picture the train, one of the ribbed stainless-steel R32s, dating to 1964, the oldest still in use. Few remain in service. Most have been "retired," meaning they left the city—they said goodbye to all that. Some have been sunk off the eastern seaboard and now sit at the bottom of the ocean, where a black sea bass weaves through the open windows and barnacles cling to what's left of the seats. A cunner fish bumps a tautog, two commuters looking for some room. Mussels crowd out a newly arrived colony of coral. The car weighs eighteen tons; it doesn't roll with the swells, but sits undisturbed, finally at rest. It is used to the darkness. The reef grows new life. One day, divers might visit.

But for now I'm still stuck slouching against the pole, staring at the drab speckled floor. A window is cracked, allowing in some stale air. Passengers shift in their gray molded seats. Someone coughs quietly into a fist. Her neighbor stares at the same folded newspaper she's been looking at but not reading for the past twenty minutes ("No Margin for Error Left, But the Mets Still Play On"). We are all thinking the same thing: something about wanting to go home, something about saying goodbye. We haven't yet realized we will be leaving part of us behind in this tunnel. We know we should be moving, but everyone just stands or sits, not talking—wondering when, if ever, we'll be coming up for air.

2014

How to Survive an Atomic Bomb

In July of 1945, the men came down from the mesa to set off a bomb. They drove for hours from the secret wartime city of Los Alamos, New Mexico, to a desert test site called Trinity that they had been readying for months. They came in cars, in trucks, in busses, in Packard limousines—an inconspicuous convoy winding south into the wastelands of the Jornada del Muerto, a dry and deadly shortcut on the ancient Camino Real de Tierra Adentro, or "royal road of the interior land," that had stretched north from Mexico City for centuries, carrying Indians, Mexicans, and conquistadors along one of the longest and oldest trails in North America that since 1680—when some six hundred colonists and converted Pueblos died in the crossing—had lived up to its name: "Journey of the Dead."

The scientists were supposed to use Americanized code names, but they were properly known as Doctors Fermi, Segrè, Bohr (father and son), Weisskopf, Ulam, Teller, Rabi, Bethe, and von Neumann, among many others—an international mix of past and future Nobel Laureates; refugees and recent immigrants from Italy, Austria, Poland, Hungary, Denmark, Germany, Russia, and such; with a pair of Americans to lead them: J. Robert Oppenheimer and Major General Leslie R. Groves. With his crumpled brown hat, Oppenheimer would become the iconic atomic cowboy, a New Yorker turned California physicist who had chosen Los Alamos as a secret weapons lab because it was near the ranchland he had come to love. The army had converted a former boys' school into barracks for the scientists; it quickly grew into a town. Accommodations were spartan, but the intellectual atmosphere was heady, electric—in some ways, they were still a bunch of boys at play. The scientists mocked the heavyset, bullheaded General Groves, who obsessed over security and—before taking control of the country's brand-new nuclear program—had supervised the construction of the Pentagon.

For years they had worked tirelessly in fear of Hitler getting the bomb; nuclear fission had been discovered in Nazi Germany, after all. (That German atomic efforts were flailing was only discovered after the war.) The best minds of Britain and Canada had chipped in. Shipments of uranium and plutonium arrived from secret plants in Washington state and Tennessee. American Counter-Intelligence Corps dogged the scientists and tried to impose discretion: nuclear fission was to be called "urchin fashion," while the bomb was known as "the Gadget."

H.G. Wells's 1914 sci-fi novel *The World Set Free* features the first use of the phrase "atomic bomb." Envisioning a weapon powered by a radium-like element, Wells writes, "These atomic bombs, which science burst upon the world that night, were strange even to the men who used them." In 1932, the book found its way into the hands of Hungarian physicist Leó Szilárd, who shortly thereafter conceived of a nuclear chain reaction. Szilárd would become one of the Manhattan Project's atomic architects, though he always claimed Wells to be the true father of nuclear weapons. In the book's dystopian future—1959—the world's cities are leveled by atomic war. One character believes that the bombs might "shatter every relationship and institution of mankind."

I fly to New Mexico on April Fools' Day, 2015. I am coming to tour the Trinity site, which the army only opens to the public one or two days a year. This July will mark the seventieth anniversary of the first detonation of an atomic bomb. The blast area—which remains radioactive—occupies an abandoned section of a massive and still very busy missile range.

Descending through choppy air into the Albuquerque International "Sunport," I look across the expanse of the adjacent Kirtland Air Force Base. I find the startling green patchwork of the base golf course and—just to the southwest—the pentagonal swath of desert that marks the Underground Munitions Maintenance and Storage Complex, the nation's largest nuclear-weapons depot, whose blast doors alone cost $7 million. Thanks to the Kirtland base, New Mexico houses more nukes than any other state. (After failing a security inspection, the unit caring for those weapons was decertified for part of 2010.) The doors to the underground bunker aren't visible from this height, but I can picture the twin ramps—one in, one out—that disappear into the earth, where some two thousand warheads wait underground.

In fission, the nucleus of an atom splits into two smaller parts, releasing subatomic particles and a burst of energy. In a hypothetical chain reaction, one neutron (supplied from the outside) splits an atom, which might emit two neutrons, which then go on to split two more atoms, which then release a total of four neutrons aimed at four more atoms, which release eight neutrons, and so on. If enough fissile material is present—called a critical mass—the reaction can become self-sustaining. The chain grows independently, exponentially, liberating greater and greater sums of energy. If there is enough mass—a supercritical amount—boom!

Criticality is a matter of density and volatility. Imagine the difference between shoving someone in an open field versus on a crowded New York subway. Some of the most dangerous experiments were ones in which the scientists tiptoed right up to that critical threshold—carefully bringing more and more fissile materials together to see when the burst occurred. They called the process "tickling the dragon's tail."

Of course there were accidents. Scientists suffered explosions and burns and berylliosis, an incurable lung disease caused by the inhalation of beryllium dust. A family cat lost its hair and died. If plutonium were to infect an open wound, protocol called for "immediate high amputation." One physicist leaned over a naked pile of uranium-235—nicknamed "Lady Godiva"—and two seconds were all it took for his body to bounce back some of the neutrons, causing the stack to begin to go critical. Noting the wild behavior of his instruments, the scientist scattered the uranium, stopping the reaction. Another two seconds would have been fatal.

The key to an explosive chain reaction is speed—the fissile material needs to go quickly from a subcritical to a supercritical mass without lingering at the threshold, which might blow the bomb prematurely (i.e., produce a "fizzle"). In 1945, atomic bombs came in two kinds. The first was a gun-type assembly, in which a cordite explosion would fire rings of enriched uranium onto a smaller uranium plug, thereby achieving a critical mass. Because the device was so simple—and highly enriched uranium so scarce—this bomb (later known as "Little Boy") could be dropped without ever being tested.

Meanwhile, the genius of the Gadget was implosion. In the second—more powerful and complex—design, a series of explosions would compress a core of plutonium into a critical state. Thirty-two precision-shaped charges (called lenses) were arranged in panels around a giant sphere (twelve pentagons and twenty hexagons, just like a soccer ball). All thirty-two lenses needed to fire simultaneously, producing shock waves that, as they traveled inward, would overlap to form a perfectly symmetrical spherical force, which would uniformly compress the plutonium into a dense, supercritical core. This bigger bomb, which would be tested at Trinity, would be named "Fat Man," though—because of the implosion method—it initially was called "the Introvert."

Some sixteen thousand nuclear weapons are sitting on the earth. Nine countries have the bomb: the United States, Russia, China, the United Kingdom, France, India, Pakistan, North Korea, and Israel. Russia and the United States account for more than 90 percent of the world's nuclear arsenal. The U.S. keeps its nuclear weapons in eleven states (New Mexico, Missouri, Washington, Georgia, California, Montana, North Dakota, Texas, Colorado, Wyoming, and Nebraska) and five European countries (Belgium, Germany, Italy, the Netherlands, and Turkey). The U.S. stockpile is estimated to be around 4,760 warheads, plus another 2,300 warheads waiting to be dismantled.

To ensure its nuclear capabilities would be impossible to destroy in a preemptive first strike, the United States maintains a holy "nuclear triad"—a strategic mix of intercontinental ballistic missiles (ICBMs), submarine-launched ballistic missiles, and nuclear heavy bombers that remain at the ready. This is deterrence. Counting nukes can be something of a shell game, but according to numbers released by the government at the beginning of the year, the triad includes 447 Minuteman III ICBMs standing in silos across the heartland, 260 Trident II ballistic missiles (which can carry multiple warheads apiece) riding on 14 nuclear-powered *Ohio*-class submarines (10 of which are usually at sea at a time), and 87 B-52H Stratofortress and Stealth heavy bombers based in North Dakota, Missouri, and Louisiana—for a total of 794 deployed "delivery vehicles" carrying 1,642 warheads. The Russian Federation deploys only 528 delivery vehicles, but—in what feels like a symbolic bit of brinkmanship—counts a single warhead more.

The New Strategic Arms Reduction Treaty, which went into effect in 2011 between the U.S. and Russia, only limits the number of deployed strategic nuclear weapons. Both sides also have "tactical"—or battlefield—nukes, plus large strategic stockpiles, as well as the material to build more bombs. The U.S. currently stores enough plutonium to make ten thousand nuclear weapons, plus the highly enriched uranium for some sixteen thousand more.

Tumbleweed blows across the parking lot of the National Museum of Nuclear Science and History (just look for the towering red, white, and blue Redstone missile—marked U.S.A.—standing by the roadside). The entrance is guarded by twin surface-to-air missiles; a giant model of an atom hangs from the façade. Inside, a bunch of kids rush toward the bathroom to wash paint off their fingers; a brochure announces the museum offers summer camp. Inside, the atomic story is told in objects: bulky machines from Tennessee used to process uranium. The custom army-green 1942 Packard limousine that ferried scientists and officers down from Los Alamos, plus the American flag that greeted them at the Trinity base camp. A seismograph used in the Trinity test. Green chunks of sand—called "Trinitite"—fused by the blast. Huge weapon casings for Fat Man and Little Boy next to models of the B-29s that dropped them. A mock-up of the Gadget itself, wrapped in wires. A photo from Hiroshima of a dead three-year-old's tricycle. A charred license plate from Nagasaki. A fallout shelter stocked with cans of General Mills "multi-purpose food" next to a Lance tactical ballistic missile bearing a W70 warhead. A Genie air-to-air rocket. A B61 gravity bomb, complete with parachute. A thermal battery, a nose fuse, a ready-safe switch. Dented bomb casings from two thermonuclear broken arrows that fell over Spain in 1966. Nearby, the Davy Crockett three-man nuclear bazooka. The Honest John rocket. A backpack nuke.

A visitor describes the concept of MAD—Mutually Assured Destruction—to his grandson. "And then we fire our missiles, and then they fire all their missiles. . . ." The kid's father walks up. The old man says, "I'm trying to explain it all to the boy." A nearby poster talks about "a stable but tense peace."

The U.S. nuclear stockpile peaked in 1967 at 31,255 warheads. We have far fewer now, but that doesn't mean our war power is impaired—one only needs so many bombs to lay one's enemy to waste, particularly when each modern thermonuclear warhead might produce ten to fifty times the yield of the Gadget. Even though the Cold War is theoretically over and the geopolitical reality has shifted away from two superpowers straddling the globe, East versus West, we are in no rush to get rid of our nukes. In fact, our arsenal seems to be leveling out.

And here's a startling fact: after adjusting for inflation, we spend more on nuclear weapons today than we did during the Cold War.

The Gadget had to be tested. A fizzle would give away the element of surprise, not to mention scatter design secrets and precious plutonium over enemy territory. General Groves needed a flat, isolated area with favorable weather that wasn't far from Los Alamos. Over locations in California, Texas, and Colorado, he chose a site to the south in the high New Mexican desert between the San Andres and San Mateos mountains, just west of the Oscura peaks. The Jornada del Muerto land was mainly owned by the state and leased to ranchers; after Pearl Harbor, much of it had become part of the Alamogordo Bombing Range. Scouting the area in September 1944, the search team was nearly bombed by friendly fire from B-17s. Construction of the camp began in November.

Oppenheimer named his test "Trinity." Even he wasn't sure exactly why, but years later, in a letter to Groves, he cited poems by John Donne that contained the lines "Batter my heart, three person'd God" and "So death doth touch the Resurrection." He had been introduced to Donne by Jean Tatlock, an old lover and Communist sympathizer who killed herself a year before the test.

The war in Europe was winding down, but the bomb project had its own momentum. Hitler was dead. So was Mussolini. Berlin had fallen. Auschwitz, Buchenwald, Bergen-Belsen, and Dachau had been liberated. Germany surrendered on May 7, 1945, the day the Trinity scientists conducted a preliminary test detonation of one hundred tons of TNT to which were added some radioactive material that would be scattered by the wind. The dirty bomb was meant to test blast effects and fallout, but the spectacular explosion—the largest to date, which was seen sixty miles away—was ultimately pointless in terms of predicting the performance of the Gadget, which would approach the TNT equivalent of twenty thousand tons. The question of radioactive fallout would rest with the physicians, not the physicists, who were mainly concerned with getting the bomb to work. General Groves was angry when he received the doctors' report—he barked at the medical director, "What are you, some kind of Hearst propagandist?" Ultimately, fallout would depend on two unknowns: the strength of the blast and the force and direction of the day's winds. The race for the bomb—that had now become one-sided—never slowed down.

The physicists weren't without worry; they theorized calamities both mundane and spectacular, from earthquakes shaking far-flung cities to the birth of a new star. One early fear involved setting fire to the earth's atmosphere. In 1942, when scientists first raised this possibility, Oppenheimer rushed to meet with his superior at the time, Arthur Holly Compton, who was at a summer cottage in Michigan. Compton drew a line. If the chances of apocalypse were more than three in a million, he would scuttle the bomb project. The scientists came back with odds just below that, so Compton decided it was worth the risk.

Many senior scientists were skeptical the bomb would even work. One hundred and two people paid a dollar to enter the betting pool to predict the force of the blast. Oppenheimer chose a mere three hundred tons. The winner—who still underbid—came closest with a guess of eighteen hundred.

Uranium for the project came from Mallinckrodt Chemical Works of St. Louis, which would go on to produce nuclear material for decades.

Radioactive waste from the company was stored around town, contaminating the banks of the Missouri and Mississippi rivers, local landfills, and a suburban creek that ran through the childhood backyard of a friend of mine, whose old neighborhood has been found to have significantly higher rates of cancer. (In 2015, the Army Corps of Engineers found radioactive soil in parks periodically flooded by the creek.) At one radioactive landfill on the other side of the airport, an underground fire has been burning since 2010—with no sign of stopping. If the fire reaches the nuclear waste, the smoke plume could rain fallout over the region.

My own university is complicit, thick with atomic ties. In 1942, the U.S. government commandeered the highly efficient one-hundred-ton cyclotron at Washington University in St. Louis—a high-tech "atom smasher" built two years earlier at a cost of $100,000—to produce microscopic amounts of the new element plutonium, thus accelerating the bomb's delivery by many months. For the next two years, in an underground building not far from my office, scientists irradiated material that was shipped to the Manhattan Project's Metallurgical Laboratory in Chicago. Thirty-one faculty members and students participated in the bomb project, many of whom would leave to do wartime work at Los Alamos. The cyclotron was overhauled in the 1960s, but the original building remains.

From 1946 to 1953, the ninth chancellor of Washington University was Arthur Holly Compton, the apocalyptic oddsmaker who encouraged Oppenheimer to press on during the bomb's early days. Compton was a deeply religious physicist who had directed Chicago's Metallurgical Laboratory, overseeing the birth of the first nuclear reactor and, later, the use of such reactors to produce plutonium (for use in the Nagasaki bomb). For three years in the 1920s, before moving to Chicago, he had chaired the Physics Department, where he made a discovery about the scattering effect in X-rays that won him the 1927 Nobel Prize. After the war, he returned to St. Louis to become chancellor, and his inauguration featured a record gathering of atomic scientists.

Compton was on the four-person scientific panel that recommended to the Secretary of War and President Truman that the U.S. drop the bomb on Japan, a difficult decision he never backed away from, citing his belief that the bomb had saved lives and might eventually lead to the obsolescence of war. On campus, his signature can be seen everywhere. Upon becoming chancellor, he recruited six chemists from Los Alamos to modernize the Chemistry Department, where their portraits still hang. A building named after him houses the physics library. He designed speed bumps of ingenious pitch—that would discourage speeding but not break any milk bottles—which I drive or walk over every morning.

In the university archives, I pore over his papers. Correspondence with atomic giants, some of which was deemed confidential, vital to the national defense, and stamped with the Espionage Act. Petitions—from scientists at Los Alamos, Oak Ridge, and Philadelphia, from the Federation of Atomic Scientists, from the students of Bennington College—calling for an international body to control the atomic bomb. (Eventually, the Atomic Energy Act of 1946 gave authority over U.S. nuclear matters to a civilian—not military—domestic agency.) An address delivered two months after Hiroshima in which Compton envisioned a global nuclear holocaust unfolding in the year 1970. A contract for $1,000 for the movie rights to his life. A stub for a $750 check for consulting services provided to St. Louis's Monsanto Chemical Company, which operated the secret nuclear laboratory in Oak Ridge, Tennessee. A letter dated October 1, 1945, from Compton's older brother, Karl, president of MIT and scientific advisor to General MacArthur, cheering for the atomic bombing of Japan. And, from Compton's personal library, Hermann Hagedorn's epic poem, *The Bomb That Fell on America*, which holds between its pages a panel clipped from a 1948 Captain Marvel comic in which the red-suited superhero battles a radioactive robot called Mr. Atom. The panel contains the story's clincher, which is bracketed in pen: "He who lives by the atom, dies by the atom."

At the Nuclear Science and History museum, slightly unnerving explosions are coming from the children's area, called "Little Albert's Lab." (It turns out the blasts are from the water electrolysis demonstration.) A medical bay shows the beneficial uses of radiation. We didn't always fear this stuff—in the 1920s, the recommended dose from a radium-lined "Revigator" ceramic water jug was six glasses a day.

An exhibit on New Mexico's Waste Isolation Pilot Plant contains cross-sections of drums containing fake radioactive waste. The nation's only long-term storage crypt for nuclear waste, WIPP stands twenty-six miles east of Carlsbad, in the southeast corner of the state. Four shafts sink nearly half a mile into an extensive mine hollowed out of a 250-million-year-old salt bed below the Chihuahuan Desert. WIPP buries nuclear-weapon waste, not material from commercial nuclear power plants. Imagine rooms the size of football fields stacked with drums of contaminated tools, clothing, soil, and such, as well as plutonium once meant for bombs.

The Environmental Protection Agency certified that WIPP would last at least ten thousand years, as the salt beds would eventually collapse, trapping the waste. The time frame seems rather arbitrary, given that the most common isotope of plutonium lasts more than twice that long. Since 1999, WIPP has entombed nearly twelve thousand shipments, totaling ninety-one thousand cubic meters of waste, at a cost north of $7.5 billion.

In early February of 2014, a salt truck caught fire in part of the mine, which was closed. On Valentine's Day, in a separate storage location, a fifty-five-gallon drum of waste from Los Alamos spontaneously ruptured, reaching a temperature of sixteen hundred degrees and leaking radioactive material that blew into the ventilation system, spread through the empty tunnels, and eventually reached the surface, where alpha radiation was detected more than half a mile away. Twenty-two workers tested positive for contamination. The likely culprit: a chemical reaction between nitrate salts and the sWheat Scoop natural kitty litter in which the waste was packed. The litter should have been clay based—in order to remain inert—but Los Alamos technicians confused the phrases "inorganic" and "an organic" when updating their packaging manual, and so a wheat-based litter was used. The Department of Energy agreed to pay the state $73 million in fines. The site isn't expected to be fully operational again until 2021, after an estimated half-billion-dollar cleanup. (Meanwhile, the long-term price tag of the accident is expected to exceed $2 billion—rivaling the amount spent cleaning up the partial meltdown at Pennsylvania's Three Mile Island nuclear power plant in 1979.)

WIPP holds only "transuranic" waste, meaning material with an atomic number higher than uranium. Thus when it is completed and sealed, WIPP will become a monument to man-made poison, a time capsule of our own folly. A series of redundant warnings is meant to remain legible for ten thousand years: giant earthen berms, massive granite monuments, buried markers, engravings, and pictographs. A 1993 report from the Sandia National Laboratories titled "Expert Judgment on Markers to Deter Inadvertent Human Intrusion into the Waste Isolation Pilot Plant" claims that no lasting barrier is feasible; there is no way to keep the future out, even once the site is sealed. The report suggests building menacing earthworks—fields of spikes and irregular blocks—around an empty center to suggest that there is nothing consecrated here, no treasure to be taken. The report proposes the use of facial icons that are not merely pained, panicked, and nauseated, but also mournful, bitter, and woeful. There is a tension inherent in the endeavor, a mix of pride and shame: while the site should be understood to be worthless, the markers must also inspire awe—or else they might be overlooked, vandalized, or erased. Ultimately, the report envisions WIPP not as a mine, or a monument, but a message—a system of meaning pantomimed across time, a history never forgotten, a past that must intrude eternally into the present:

"Sending this message was important to us. We considered ourselves to be a powerful culture."

"This place is not a place of honor."

"What is here is dangerous and repulsive to us."
"The danger is still present, in your time, as it was in ours."
"The danger is to the body, and it can kill."

The month I was in New Mexico, the governor wrote to the U.S. Secretary of Energy, asking that an area not far from WIPP be considered as a repository for spent fuel rods from nuclear power plants—the most dangerous radioactive waste in the country.

At the Trinity site, the men watched 16 mm movies—*The Prisoner of Zenda* and *Beau Geste*—under the stars. Soldiers and scientists played volleyball, poker, and polo (on horseback with brooms and a soccer ball). With submachine guns, they hunted antelope, which were served up as steak. With little else around them but aerial targets, one night the base camp was bombed in error by planes from the nearby Alamogordo Bombing Range, setting the stables afire. Three days later, the carpentry shop was hit. The test director suggested installing anti-aircraft guns.

By July 1945, the month of the test, nearly three hundred people were housed at Trinity, a number that would rise to 425 two weeks later for the test weekend. A laboratory now occupied the alkali plain, which was crisscrossed with five hundred miles of instrument lines stretched between wooden T-poles standing about the height of a man. More wires were buried in garden hoses in the sand. There were impulse meters, geophones, and peak pressure gauges. More than fifty cameras—motion-picture and still—were assembled to capture the explosion, everything from high-tech Fastex devices recording ten thousand frames per second behind bunkers of steel and glass to a simple pinhole camera, operating under the ancient principles of the camera obscura. Photographers were given incomplete information, told simply to be ready to film something historic that would begin with the light of ten burning suns. (The only successful color photo of the blast would be taken by a technician who—as an amateur photographer—had brought his own camera.) Three shelters with concrete slab roofs were buried ten thousand yards to the north, west, and south of Ground Zero, where a one-hundred-foot steel tower rose above the desert. At the top stood an oak platform shielded by corrugated iron. The tower was anchored by concrete footings buried twenty-five feet deep.

The test was originally targeted for July 4, but by mid-June Oppenheimer said the thirteenth was the earliest it could be. Eventually, a date came down from Groves—July 16, during the predawn hours—despite that it was not within the first or the second range of days given by the chief meteorologist, but instead would fall during a period of predicted storms, which could concentrate fallout.

At the museum, I step outside into Heritage Park, where the thirty-foot sail of a nuclear sub rises improbably out of the sand. A B-29 bomber looms beside the atomic cannon; a sign warns KEEP OFF. A military plane drones low overhead, coming in for a landing at Kirtland, and beyond the fence I see the glass-faced tower of the Sandia National Laboratories complex, one of the country's three remaining nuclear labs, where the parking lot is full. Behind a B-52 stands an enormous Mark 17 bomb, just like the one that on May 22, 1957, accidentally slipped out of the bay of a bomber landing at Kirtland. The 41,400-pound thermonuke—the largest the U.S. ever deployed—fell just outside of Albuquerque on land owned by the University of New Mexico, blowing a twenty-five-foot crater and killing a cow. Had the bomb been armed, the explosion would have been five hundred times that of the Trinity blast.

On the other side of the park, a rusting three-stage Minuteman missile lies along the fence, its sign also down, having fallen against the tail. I turn, and the wing of a B-47 bomber almost clips my brow. I put my head into the blackened thruster of a colossal Titan II intercontinental ballistic missile lying in stages on the ground and imagine the heat, smoke, and smell of all that it burned.

Four days before the test, on Thursday, July 12, the massive body of the bomb—the high-explosive sphere that would encircle the core—was bagged in plastic, boxed in pine, covered by tarp, and tied to the back to a five-ton truck, which left the mesa at midnight, just as the calendar turned to Friday the 13th. The scientists laughed at superstition, but two Army Intelligence cars joined the convoy, which blared a siren as it rolled through sleeping Santa Fe, to discourage drunks from crashing into them.

Earlier that day, an army sedan had left Los Alamos carrying physicist Philip Morrison, who, in a month, would assemble the bomb dropped on Nagasaki, before becoming a popular scientist and staunch opponent of nuclear proliferation. Morrison sat in the backseat next to a shockproof box—designed to bounce if dropped—that contained the two hemispheres of the Gadget's plutonium core. Upon arriving at the Trinity site, the army lieutenant was directed to deliver the top-secret package to a ranch house two miles from Ground Zero.

On the morning of Friday the 13th, in the master bedroom of the ranch house, where the windows were taped against the dust, Canadian physicist Louis Slotin began assembling the bomb's plutonium core, which was about the size of an orange but weighed a startling thirteen pounds. Slotin Scotch-taped a small neutron initiator made out of polonium and beryllium—called an "urchin"—into a hollow pit in the middle of the plutonium hemispheres. Upon being crushed by the shock wave of the explosions, the urchin would release a burst of neutrons to start the chain reaction. The plutonium sphere itself would nestle inside an eighty-pound cylindrical plug made of uranium-238, part of the "tamper" that would reflect more neutrons back onto the critical mass and inhibit early expansion of the material, which might weaken the reaction.

Slotin would earn the nickname "chief armorer of the United States." Ten months later, he would be killed after showing some colleagues—on a whim—how to tickle the dragon, lowering the top half of a tamper onto a plutonium core using nothing but a flathead screwdriver to hold the hemispheres apart. The screwdriver slipped, the sphere became whole, and there was heat and a bright-blue flash. Slotin flung off the tamper, but it was too late—he soon began vomiting. (Three of the other seven witnesses would later die from radiation-related diseases.) The core had already killed Slotin's assistant nine months earlier, when he dropped a tungsten brick next to it. He had taken about a month to die. Slotin would last nine days, after his parents had flown in, his face turned red, his skin blistered, and his hair began falling out. Coughing but coherent, he suffered through his final hours struggling to answer questions on camera. The footage was made into a training film for nuclear technicians that was shown at least until the late 1970s.

At 3:18 p.m., the Trinity core arrived at the base of the tower, where it was to be inserted into the two-ton, five-foot ball of explosives. The body of the bomb was a carefully calibrated device: the implosion lenses had been X-rayed, and each tiny air pocket had been drilled and filled with explosives. The blast waves had to arrive in precise synchronicity. Some of the charges were made to fit snugly with tissue and Scotch tape. A declassified film shows the scientists working under a large canvas tent. Oppenheimer leans over his Gadget, hat on, sleeves rolled up. The men—some topless, others in T-shirts and tanks—stick their arms deep into the top of the bomb. There was a moment of panic when the cylindrical plug—containing the plutonium core—didn't fit, but someone realized the metal had simply expanded in the heat. Once it cooled off, it slid in perfectly. One step at a time, the scientists followed the painstaking assembly instructions set down for the "hot run": "Place hypodermic needle *in right place.* (Note: Check this carefully) . . ." and so on. When they were done, some men went swimming in a water tank. The next morning, a $20,000 winch raised the Gadget to the top of the tower, where the detonators were installed. As it went up, a pile of striped G.I. mattresses were placed beneath the bomb. A handwritten sign advised WEAR A HARD HAT.

That night in Albuquerque, I dream of bombs going off. In the morning, I wake to hear Russia is threatening to use nukes over the Ukraine. Later, over huevos rancheros, I read that a break has come in nuclear talks with Iran. Driving south out of town, I see the ballpark for Albuquerque's minor-league team, the Isotopes. Winds whip across the desert, throwing a mattress off the roof of a car and onto the highway. I keep the dark mountains on my left as dust storms roll through the valley like fog. I pass a dead coyote on the road.

We are facing a future of fewer but better bombs. The average U.S. warhead is twenty-seven years old. The government expects to spend more than $350 billion in the next decade—and at least $1 trillion in the next thirty years—to update its nuclear arsenal. Plans include designing a new nuclear sub, a long-range bomber, and an air-launched cruise missile, as well as looking into the next generation of land-based ICBMs. Older designs will be retrofitted. One bomb—the B61, which dates to 1963—is scheduled to get a new tail assembly, making it less of a gravity bomb and more of a guided weapon, greatly increasing its accuracy (and lessening the need for a huge yield). The B61 upgrade will cost more than $10 billion, which—as the *Bulletin of the Atomic Scientists* has pointed out—means each new bomb will cost more than if it were made of gold. Waste is one thing, but analysts have noted an even more troubling drawback: with more precision—and fewer unintended casualties—comes a greater enticement to actually use a nuke.

I spend the night in Truth or Consequences (pop. 6,100). A spa town, "T or C" originally was incorporated as Hot Springs, but, in a 1950 publicity stunt, renamed itself after a popular quiz show. After a multimillion-dollar commercial "spaceport" opened nearby in 2011, the town has geared up for an influx of space tourists that may or may not ever arrive. (Flights, on carriers such as Virgin Galactic, have yet to take place.) I'm staying at the Rocket Inn—in the Apollo Room, as it happens, which celebrates our mission to the moon, which blazes nearly full tonight, ringed by a double halo. As I stand ogling it, one of the motel owners strolls by, saying, "Wow, look at that." The air smells of wood smoke. I go back inside and prowl around my room. The hot springs that bubble up into the historic bathhouses downtown are heated by the same radioactive decay that fires the inner earth. Last night, in my atomic nightmares, I was stranded between terror and wonder—panicked, but giddy to be at Ground Zero. How I wanted to see the blast, to witness the fallout. Tonight is so quiet. You can hear every stray voice or dog, every screen door slam, every car hit the gas as it decides to keep on rolling through the valley. I know the name is just a worn-out gimmick—one that doesn't ring many bells anymore—but tonight the choice feels dire, fatalistic. Truth or Consequences. Well, which is it going to be?

Everyone pays lip service to a world without nukes, but we are not the only ones refusing to put away our bombs. Russia is upgrading Soviet-era systems, enhancing its bomber force, and building three new land-based missiles. Meanwhile, China is adding to its arsenal, while India and Pakistan are increasing their abilities to produce plutonium and uranium. Britain is building new missile subs, and France and Israel are improving their capabilities. Meanwhile, as global disarmament stalls, the nuclear have-nots express growing resentment.

We have nonnuclear weapons of amazing power and precision—the technology of conventional war has raced ahead, unchecked—but the point of a nuclear weapon is fear: the fear that someone, somewhere (either us or them) will get pushed too far, and everything—*everything!*—will come to an end. Mutually Assured Destruction is a position of pure reason and pure madness, and it's not a line to be crossed, or a point to tip past, but a delusional dream state that we've inhabited for more than sixty years. The logic of deterrence is a closed circle, a serpent eating its own tail. There will always be another enemy—monolithic, inscrutable, duplicitous—that must be kept cowed.

There is pleasure and comfort in that. But, for example, what use is a nuke against a man with a box cutter?

Obsolescence takes a mental toll. For instance, missileers know that owing to geography, in order to fire our ICBMs at China, Iran, or North Korea—today's lineup of "villains"—the missiles would have to fly over Russia without provoking a response—a wildly unlikely scenario, to be sure. America's nuclear morale—on bases, in labs and facilities—is dismal. Ambitious young scientists aren't eager to babysit old bombs that can no longer legally be tested. Soldiers don't want to sign up to sit in useless silos. Stories of drug use, security breaches, boredom, burnout, cheating, and catastrophic close calls have appeared recently with some regularity—including the one about the six thermonuclear missiles that were "lost" for thirty-six hours and mistakenly flown across the country, unguarded. Or the eighty-two-year-old nun who managed to breach the nation's most secure uranium storage facility deep in the Tennessee woods, where she and her two companions splashed blood on the building.

Late last year, even the Vatican, long a believer in deterrence, backtracked on nukes, calling the very possession of them immoral.

Policy-wise, the U.S. occupies a weird middle ground—we want to manage a reduced but still-robust nuclear stockpile, while somehow imparting the idea to our nonnuclear allies and enemies that atomic weapons aren't the way to go. The official government rhetoric walks a curious line, downplaying our commitment to nukes while at the same time confirming our willingness to use them, if pushed. Meanwhile, we're all up to our old Cold War stunts again, Russia buzzing Europe and Alaska with its nuclear bombers, while we test-launch two ICBMs in a single week.

Atomic tourists aren't the only ones making pilgrimages to New Mexico. The state also advertises a fifty-two-stop "space trail," which includes a visit to Spaceport America, which was built in one of the poorest states in the country with some $200 million of New Mexican taxpayers' money. The next morning, I get up early for a tour. I am the last of the fifteen space enthusiasts in the van. Our driver, Gary, talks nonstop over the PA during the forty-five-minute trip. We pass the beautiful Elephant Butte Dam, part of the Rio Grande Project, which delivers water to New Mexico and Texas. I am startled to see a sparkling blue expanse stretching miles into the desert, the reservoir where the frazzled Trinity test director and his assistant recovered on a fishing trip before returning to dig out their equipment. Everyone in this part of the state talks about water.

Gary asks for the hand of anyone who has signed up for a Virgin Galactic flight. No one moves; we are strictly terrestrial space tourists. He says, "I understand. It's that big number with a lot of zeros between us and space." Specifically, a flight costs $250,000, a price some seven hundred would-be astronauts have already paid. Buying a ticket is hardly a sure thing. Even before the October 2014 crash that delayed Virgin Galactic's schedule for luxury space flights, the spaceport was sitting empty, mainly used for photo shoots by Land Rover and Kawasaki.

Why build the world's first spaceport in the middle of nowhere, New Mexico? The reasons are surprisingly similar to the Trinity test. First, Gary tells us, it has to do with seclusion and privacy. "There are more cows than people. A lot of companies don't want the world looking over their shoulder." Then there's the weather: an average of twenty-eight sunny days a month. The high elevation is also a boon—"we say you get the first mile for free!" And thanks to the adjacent White Sands Missile Range—the vast army installation that has swallowed up the old Alamogordo Bombing Range and the Trinity test site—the airspace above this area is restricted upward to infinity. Gary tells us the White House is the only other location with complete overhead clearance, before launching into a long argument for privatized space. Meanwhile, we pass a dirt road tentatively scheduled to be paved by the end of the year, which would connect to the interstate and thus bypass Truth or Consequences.

The spaceport's Gateway to Space Building nestles between two earthen berms. From the back, the oxidized-steel hangar looks like twin turtle shells. It's off-limits to tourists, but I imagine the rich and famous disappearing into the darkened entryway, where they will board a ship that will allow them to slip the bonds of earth—for only a few weightless minutes—after which they, like everyone else, will be brought right back down where they came from.

I drive north to Socorro. Colorful crosses hung with paper flowers line the side of the highway. More death in the desert—I'm reminded that today is Good Friday. The sun is bright; I keep seeing shapes in the brush and mirages on the road. At last I pull into the old town square of Socorro, where men sit on benches in the shade and a historic marker talks of what happened at Trinity seventy years ago. Tomorrow the site will be open to the public. As I check in to my hotel, I hear British accents behind me.

At the northern edge of White Sands, a notice says the road can be closed for missile firing and gives a number to call. Seventeen miles past the gate to the range—which won't be open until 8:00 a.m. tomorrow—a tall, weathered wooden sign—ROCK SHOP—springs out of nowhere. I pull over and two big dogs start going nuts behind a fence. The yard is strewn with metal tables piled with colorful slabs, geodes, agates, jaspers, boulders, and, strangely, seashells. A woman in a tank top appears and asks what I want. I tell her I'd like to see the rocks. She unlocks the shop, which has been open since 1968 and consists of a number of rooms stuffed with rocks, minerals, gems, and jewelry. The owner, a suntanned older woman with very blue eyes, says she took over in the mid-1990s.

I buy a small piece of Trinitite that was picked up by a miner before the blast field was bulldozed in the 1950s. The glazed green rock is illegal to collect, but my piece is grandfathered in, she tells me. She says it emits .08 millirem per hour of alpha radiation. (A Geiger counter will measure somewhat less than that back in St. Louis.) I ask if the rock is dangerous. "Just don't crush it up and snort it," she says. And here she's not joking. While alpha particles usually can't penetrate the skin, they are bad news inside the body. She doesn't mention the beta particles, which I know the piece is sending out, too, though they're also relatively weak and stopped by a thin layer of metal.

What was borne in that green glass? A mottled chunk of plutonium, uranium, sand, steel, minerals, and dust, forged in fire and rained down from above, the shiny side cooling to reflect the sky. I'm not sure why I buy the sample. Partly out of guilt for making the woman open the shop just for me—the last visitor signed the guest book a month ago—but that's not entirely it. I'm standing about 18.5 miles to the northeast of Zero. This tiny town—now essentially gone—was bathed in radiation that day. The readings were some of the highest measured. As I drive away, I will think of roving monitor Arthur Breslow, who—chasing the cloud—left his respirator at a searchlight station he had been forced to evacuate, and so drove on through the radioactive valley with his windows rolled up, breathing through a piece of bread. One family who lived off this road was advised to stay indoors—for days. This hot rock is a terrible memento mori. I put it far away on the floor of the passenger seat. The last thing the woman at the shop told me: "Tomorrow is going to be a wild day."

The scientists' to-do list for Sunday, July 15, the day before the test, included "look for rabbits' feet and four-leaved clovers." Late that night, a thunderstorm moved in. The youngest scientist, Don Hornig, who had designed the electric trigger, was dispatched to the open metal shack at the top of the tower, where he sat alone with the bomb, which was now wrapped in thirty-two thick detonation cables set to simultaneously explode. Everyone was worried about sabotage and lightning. A week before, static electricity had prematurely triggered the firing unit. Hornig was unarmed and unclear what to do in the case of an emergency. He passed the time reading humorous essays by the light of a dangling bulb. Later, he would advise Eisenhower, Kennedy, and LBJ, and become the president of Brown University. He was the last to leave the tower.

The scientists were exhausted and on edge. Four psychiatrists from the secret site in Tennessee were ready to be flown in. The day before, a nonnuclear test of a dummy bomb had failed. (The lenses were not right.) The explosive expert bet Oppenheimer a month's salary that the lenses would work on the real thing. Everyone was watching the sky. Moisture might short out the triggers; storms could blow radiation over nearby towns. Surveying the Oscuras in the dusk, Oppenheimer cryptically confided to a metallurgist: "Funny how the mountains always inspire our work."

Scientists took their stations. The southern shelter was the control room. The other two housed searchlights, cameras, and instruments. More scientists would observe from base camp, ten miles to the southwest of the tower. The VIP area—which would include nonessential staff, visiting scientists, military men, and a single journalist handpicked from the *New York Times*—was on a hill about twenty miles to the northwest.

At 1:00 a.m., General Groves was trying to catnap in a poorly secured tent when he was awakened by canvas flapping in the wind. At 2:00 a.m., just as the busses of VIPs were arriving from Los Alamos, more violent thunderstorms rolled into the area, lashing the control bunker and base camp with high winds, but for the most part sparing the tower. The shot was scheduled for 4:00 a.m., but at 2:00 a.m. it was decided to postpone the test until at least five o'clock, when the meteorologist thought the storms would clear. General Groves threatened to hang him if he was wrong.

At 2:45 a.m., the general called the governor of New Mexico and alerted him that it might become necessary to enforce martial law. The reporter from the *Times* had left several press releases back in New York, including one that described an accident at Oppenheimer's mountain ranch that had claimed the lives of many prominent scientists, as well as the writer's own. Base camp began serving breakfast at 3:45 a.m.: powdered eggs, French toast, and coffee. Meanwhile, two physicists observed a heap of frogs breeding noisily in a rain-flooded hole. The weatherman made his last forecast at 4:15 a.m. At eight minutes past five, the test director examined it. Both men had been awake for two days. The conditions were far from ideal, but by a stroke of luck, the winds were favorable.

Twenty-nine people were in the north shelter, thirty-seven in the west, and thirty-three in the south, the control bunker, where Oppenheimer was stationed with other crucial personnel. General Groves watched separately—at base camp—to lessen the chances of both men being killed (which would be a terrible setback for the project). In an attempt to frighten the guards, one Nobel Laureate began taking bets on whether the bomb would destroy the whole world, or merely New Mexico.

Trinity was plagued by echoes: the shortwave frequency happened to overlap with a freight yard's in San Antonio—the scientists and the rail men could hear each other's traffic. The ground-to-plane frequency matched the Voice of America's, startling some physicists with bars from the "Star-Spangled Banner." On the observation hill, theoretical physicist Edward Teller, who would become known as the "father of the hydrogen bomb," passed out suntan lotion, which everyone applied in the dark. The Gadget was twenty miles away. Teller wore heavy gloves and welder's goggles—to the dismay of some of the more unsuspecting MPs standing by. Down the road, a man named J. E. Miera, who grilled hamburgers for the scientists in his popular Owl Bar, was awakened by soldiers sitting outside with seismographs. They told him to come out front to witness "something the world has never seen."

The Owl Bar and Café still stands in the little town of San Antonio. Tonight, a tattooed older vet in a black T-shirt is drinking tequila and a beer. He used to work at Los Alamos and tells the bartender, "I don't exactly see why everyone gets so worked up about visiting where they set the bomb off." The bartender tells him about local kids who grew up with problems, special needs, cancer. Despite his nonchalance, the man says he's thinking about finding a spot and sleeping by the side of the road to avoid the line of cars in the morning. "There's nothing posted against it," he says.

Rowena Baca tells me a story she has told many others who have come through her bar, a beautiful old dive adorned with autographed photos and dollar bills pinned to the walls. She is the granddaughter of J. E. Miera, the grocer who was woken early and told to come outside, where I had just parked my car. "The men—I don't know whether they were soldiers or scientists—would rent cabins from my grandpa," she says. "They said they were prospectors. Daddy always said they were really nice guys."

Just after 5:00 a.m., the test director unlocked the switches. Soldiers stood in slit trenches. Radiation monitors swung from blimps that would be vaporized upon relaying their data. At 5:03 a.m., as the timer began, the order was repeated for everyone, everywhere: lie facedown on the ground, feet to the blast—don't look until the first flash is over. Two minutes later, the last men left the base of the tower, calmly driving five and a half miles to the southern shelter. At 5:10 a.m., over a loudspeaker, Samuel K. Allison of the University of Chicago began the first countdown in history. (He came up with the idea to count backward to the blast.) General Groves rode a jeep to base camp, where one man insisted on facing the explosion. Everyone else lay down in the trenches.

At T-minus five minutes, a green warning rocket flared. Another, three minutes later, refused to fire. With a minute to go, Oppenheimer is said to have remarked, "Lord, these affairs are hard on the heart." He gripped a post and seemed to hold his breath. At T-minus forty-five seconds, one of the arming party threw the final switch. The bomb would now fire automatically. A chime accompanied Allison on the countdown. At ten seconds, a gong clanged in the control shelter. One physicist cried to another, "Now I'm scared!" Most everyone was praying. At nine seconds, interference with a local radio station cut in over the loudspeaker—and the scientists were treated to the cheery dissonance of Tchaikovsky's "Serenade for Strings," which was currently featured in the MGM musical *Anchors Aweigh.* At base camp, General Groves lay on the ground between two scientists thinking of what he would do if the zero came without a bang.

At the Owl, Rowena Baca sits at a twenty-five-foot solid mahogany bar her grandfather installed in his store more than seven decades ago. He paid some soldiers to walk the bar three-quarters of a mile down the road from an old rooming house that belonged to Gus Hilton—of *that* Hilton family, who hail from San Antonio. It took the soldiers two days. They were paid in hamburgers and beer. Years later, Conrad Hilton's son returned, trying to buy back the bar. He promised to build her a replica. She refused. "To me, it's priceless."

The guy to my left was last here in 1962, when he got a job right out of high school working on the interstate. "We were only eighteen, nineteen, but Frank," meaning Baca's father, "would let us drink beer." He says the place has stayed the same. Baca still gets to work at 7:00 a.m.—though she closes the bar earlier these days. She says, "Daddy paid me fifteen dollars a week when I was twelve—and I've been in the grocery business ever since."

She has noticed a steady rise in tourists attending the Trinity open house. She says Japanese visitors have started to show up in recent years, too. She brings up the protesters who will come from across the state to be here tomorrow—they want recognition and compensation for being atomic test victims. Baca says, "We were the ones closest, but all the old timers are gone. There's nobody left to get the money."

Baca was a toddler that morning in 1945. She remembers, "Grandma thought it was the end of the world. Everything was red. She threw me and my cousin under the bed." Then she adds, "We were crying under there. I don't know why she thought the bed would save me."

Finally, on July 16, 1945, at 5:29:45 Mountain War Time, the Gadget exploded in the predawn dark with the light of twenty suns.

Someone said it smelled like a waterfall. But first came a flash visible from three states that dimmed to reveal a boiling mushroom cloud that shot skyward in perfect silence, burning red then luminous purple as it ionized the atmosphere. In seven minutes, the column would stretch more than seven miles tall. The MP guarding the door of the control bunker went pale; nobody had thought to explain to him what would unfold. At the site of the roiling cloud, an officer worried, "The longhairs"—or scientists—"let it get away from them," while General Groves deadpanned, "Well, there must be something in nucleonics after all."

A physicist remembered, "It was like being at the bottom of an ocean of light. We were bathed in it from all directions. The light withdrew into the bomb as if the bomb sucked it up. Then it turned purple and blue and went up and up and up." The War Department officially reported a radiance "golden, purple, violet, gray, and blue" that "lighted every peak, crevasse, and ridge of the nearby mountain range with a clarity and beauty that cannot be described." From a car headed to Albuquerque, a partially blind music student even saw the flash. Physicist I. I. Rabi on the light: "It blasted; it pounced; it bored its way right through you."

The heat struck hard in the cool morning. To observers in base camp—ten miles away—it felt like standing in front of a fireplace. A glowing yellow turbulence raced across the desert floor, whipping up sand—the shockwave beating the ground toward them. Unable to keep still, one scientist dropped six pieces of paper before, during, and after the wave's arrival—just another way of calculating the force of the blast. It took forty seconds for the shock to hit base camp, and with it finally came the sound—a deep booming that ricocheted through the valley and canyons until the echoes collapsed into a continuous roar. Thirty-four years later, one witness would write, "I can still hear it."

Outside the control bunker, the shockwave flattened the explosives expert, who had neglected to duck. He got up and hugged Oppenheimer, before asking his boss to make good on their bet. A conga line broke out, and people took turns shouting over the PA. Elsewhere, a man who had burned his corneas was given morphine.

The *New York Times* journalist would write, "One felt as though he had been privileged to witness the Birth of the World—to be present at the moment of Creation when the Lord said: 'Let There Be Light.'" Oppenheimer strutted around like a cowboy. The test director told him, "Now we are all sons of bitches." One scientist passed whiskey.

Another asked, "What have we done?"

I wake before dawn to a blood moon setting over the western mountains. An auspicious alignment: the total eclipse of a full moon, the earth's shadow casting it a fearful red—an ancient sign of apocalypse. Recently, a few fringe Christian ministers have prophesized that this blood moon—one of four in a row—foretells the end of the world. (Revelation 6:12: "And I beheld when he had opened the sixth seal, and, lo, there was a great earthquake; and the sun became black as sackcloth of hair, and the moon became as blood.") From my hotel parking lot, I stare dumbly at the red orb as a biker straps gear onto his ride. The air is cold and clear; the sun won't be up for a while.

In the lobby, a flyer advertises a religious revival called "the minor prophets' guide to the end times." It's at a place called Quemado—or burned—Lake, and promises "good old Bible preaching." Yesterday was Good Friday. Last night was Passover. A time of great Biblical death—Christ on the cross, all those firstborn Egyptian sons. Today is Holy Saturday, the day Jesus descended into hell. Tomorrow will be Easter. And so I go to Trinity.

The smoke covered the ground for an hour. When the fireball touched down, it blasted a half-mile crater. Everything within another half mile beyond that was dead—down to the ants. The smell lasted three weeks. Doors were torn off a farmhouse three miles away. Downwind, cattle would die. At twenty miles, a black cat's fur would go white. Five miles beyond that, monitors found a stunned mule, tongue lolling out. The air blast broke windows in Silver City and Gallup, the latter some two hundred miles away. In a 1965 television documentary, Oppenheimer recalled, "We knew the world would not be the same. A few people laughed, a few people cried. Most people were silent. I remembered the line from the Hindu scripture, the Bhagavad Gita. Vishnu is trying to persuade the prince that he should do his duty and, to impress him, takes on his multi-armed form and says, 'Now I am become Death, the destroyer of worlds.' I suppose we all thought that, one way or another."

Four hours after Trinity, the *Indianapolis* slipped out of the San Francisco Bay, headed to the western Pacific and carrying in its hold the Little Boy bomb that three weeks later would explode over Hiroshima.

At base camp, General Groves wanted to wait out the fallout by discussing the logistics of the next assignment—dropping a bomb on Japan—but found, to his disappointment, that the scientists weren't in the right "frame of mind." Oppenheimer's younger brother, also a physicist, described a feeling of dread: "I think the most terrifying thing was this really brilliant purple cloud, black with radioactive dust, that hung there, and you had no feeling of whether it would go up or drift towards you."

Within fifteen minutes, the mushroom cloud had divided into three, a most unholy trinity. The lowest portion moved north, while the middle went west. Some fifty thousand feet up, the largest and highest part drifted northeast—exactly as desired. Another fifteen minutes later, the top of the mushroom was said to resemble North America, while the remaining clouds formed a reddish-brown question mark. Five minutes later, the lowest, heaviest cloud swept over the north shelter, forcing it to evacuate. But for two hours, very little fallout came down, which raised hopes—before the top cloud irradiated a long stretch of land one hundred by thirty miles. (That day, gamma rays would be detected 260 miles away in Colorado.)

Twin B-29s followed the high white cloud for miles, before losing it in the thunderheads. Other aircraft tracked it for several hours beyond that. The head meteorologist, at the controls of his own plane, estimated that in thirty-six hours it would circumnavigate the globe. Before then, the first radioactive cloud slowly sailed east over Kansas, Iowa, Indiana, New York, and New England.

The protesters carry signs that read NEW MEXICO IS NOT RADIATION PROOF and SPEAKING UP FOR THOSE WHO HAVE BEEN SILENCED BY THE BOMB. After I pass them, I will spend about an hour waiting in a line of vehicles four miles long to get to the entrance of the missile range. A tiny Yorkie hangs its head out of the Jeep behind me, which makes me think of the army dogs that used to sniff out exploded missile fragments in the sand. I see cattle and smell burning rubber. A motorcycle cruises by in the left lane, cutting the line, and people honk. On the mountains ahead, I can see the telescopes of the Lincoln Near-Earth Asteroid Research project, which for decades have watched the skies for threatening space objects.

At thirty-two hundred square miles—the size of Rhode Island plus Delaware—White Sands Missile Range is the largest military installation in the U.S. Since 1945, the range has hosted a number of historic moments in rocketry, including the first stateside launch of a V-2 and the early tests of the Redstone, the first rocket to not only carry a live nuclear warhead, but—three years later—an American astronaut. A projectile from White Sands took the first photograph to reveal the earth's curvature from space. Over the years the range has tested everything from bunker busters to rocket sleds to airborne lasers. It has hosted some notable visitors, including President Kennedy (who, in 1963, watched the army fire a Little John missile),

David Bowie (who, in 1976, filmed parts of *The Man Who Fell to Earth*), and the Space Shuttle Columbia (which, in 1982, landed here because of flooding in California).

Today, a brochure from the test range details its mission: "We shake, rattle, and roll the product, roast it, freeze it, subject it to nuclear radiation, dip it in salt water, and roll it in the mud. We test its paint, bend its frame, and find out what effect its propulsion material has on flora and fauna. In the end, if it's a missile, we fire it, record its performance, and bring back the pieces for postmortem examination. All test data is reduced, and the customer receives a full report."

Later on the test morning, a lead-lined Sherman M-4 tank—complete with its own air supply—rolled into the blast zone to scoop up soil samples. Even though the ground immediately beneath the tower was paved to reduce the amount of earth picked up by the blast (which would become fallout), the fireball vaporized between 100 and 250 tons of sand, much of which rained back down as the radioactive glassy green residue dubbed Trinitite. The desert floor had been glazed for nearly half a mile. The tower was reduced to a red stain on the ground and a few fingers of rebar sticking up from its footings. The half-mile crater sloped down about ten feet—as if the earth had been pounded in, not blasted apart. When General Groves saw the hole, he was reported to have said, "Is that all?"

To preserve secrecy, the local area had not been evacuated beforehand. But food and water were stocked, and trucks and jeeps stood ready to move the population. Base camp was prepared to hold an extra 450 people. Other military installations could take in evacuees. The press was fed a cover story about the explosion of a munitions dump at Alamogordo Air Base.

Fallout monitors roved the countryside, taking readings; after one day, the cars themselves would become radioactive. The doctors were chiefly concerned about high-intensity exposure, not long-term consequences, which weren't fully understood at the time. Two towns were nearly evacuated, but levels dipped once the "hot" cloud moved on.

Grazing land on the Chupadera Mesa, thirty miles to the northeast, was particularly contaminated. Fence posts were blanched white, as was the beard of a rancher. With their thick wool, the sheep fared better than the cattle, whose fur began to fall out about a month after the blast. (It would grow back without pigment.) The government bought four of the "atomic calves," which—once Trinity's secret was revealed in the weeks following Hiroshima—became popular in the press and were displayed across the state. When tests confirmed radiation was to blame, the other scabbed and splotched cattle were purchased and sent to Los Alamos and Oak Ridge, where they were bred, studied, slaughtered, and, in some cases, eaten.

One rancher would describe his land being covered with "light snow," and a homeowner twenty miles away remembered flour-like dust remaining on the ground for four or five days. Ted Coker, who sold some of his afflicted cattle to the government, would recall the "funny" smell of the fallout; he was among a number of locals who eventually died of cancer.

Wind and rain concentrated the fallout that day. One area thought to be deserted—nicknamed "Hot Canyon" because the radiation was off the charts—was later was found to be inhabited by an older couple and their ten-year-old grandson, who drank the water that collected on their tin roof. Under various pretexts, scientists, doctors, and intelligence agents visited the family seven times in the two years after the blast. Reasons for the visits were never made explicit, even after the Trinity test became famous. The long-term health of overexposed civilians would not be monitored. Decades later, the doctor in charge of the Trinity medical group would admit, "We just assumed we got away with it."

Only 20 percent of the bomb's plutonium underwent fission; the other ten and a half pounds were pulverized and scattered over thousands of acres of desert. The half-life of plutonium is twenty-four thousand years. Inhaling only a small amount can lead to various deadly cancers.

At the zero mark, some 360 radioactive isotopes were born. Some died immediately, others lingered—a long chain of decay. Showing up fourteen minutes after the fact, strontium-90 is a soft metal with a twenty-nine-year half-life that is absorbed by plants and thus moves up the food chain, where it accumulates in bones, like calcium, and irradiates the body from within. (The isotope can move far and fast. In 1953, milk in New York was found to be tainted with strontium-90 from nuclear tests in Nevada.)

In 1947, a secret study from UCLA found yucca had recolonized the blast zone. The crater was filling in; it now dipped only six feet. But in places the Trinitite was still half an inch thick, and the soil radioactive to a depth of three and a half feet. Plutonium was on the ground for eighty-five miles. The scientists observed birds with deformed claws, rodents with cataracts, and oddly marked beetles. A study the following year found that the wind continued to spread the contamination.

For at least five years, the army would deny the existence of any fallout. Eight weeks after the blast—and a month after the bombing of Hiroshima and Nagasaki—reporters were invited to tour Ground Zero wearing little white booties to dispel the notion of lingering radiation poisoning in Japan. The journalists—from *Time* and *Life* and smaller local outlets—ate chicken and took home Trinitite souvenirs, while photographers shot pictures of Oppenheimer and Groves posing next to the remains of Ground Zero. Groves made his driver stand in the crater for half an hour to show just how harmless the site was. Two decades later, the man would be diagnosed with leukemia (which would prove fatal). The cause of cancer can be hard to determine, but before the driver died, the military granted him "service-connected" disability.

A survey in 1955 would be the last for a while. People would continue to be exposed by the gardens they grew and the animals they grazed (and watered with contaminated groundwater)—reaping their own radiant crop of milk, vegetables, chickens, goats, and cows. The government never intervened. A CDC report concluded in 2010, "It appears that internal radiation doses could have posed significant health risks for individuals exposed after the blast."

At the gate, security officers check IDs and peer into cars. A tall metal sign warns in English and Spanish: AREA MAY BE CONTAMINATED WITH EXPLOSIVE DEVICES. More protesters hoist homemade signs. Some carry the names of people—even entire families—killed by cancer. A man in a dust mask holds the phrase GONE TOO SOON while another man waves the WINDS OF DESPAIR. A poster quotes the book of Revelation: I WILL DESTROY ALL THOSE WHO DESTROYED THE EARTH. A woman in a black-and-white Day of the Dead mask brandishes a sign that reads: REMEMBER DOWNWINDERS. Through my window, I ask if I can take her picture. She says, "Please do," then shouts, "No more silence after seventy years!"

After the bombing of Japan, the U.S. government continued testing atomic weapons until 1992, setting off more than one thousand nuclear blasts, or an average of one every 16.5 days. (In 1962 alone—as a ban on atmospheric testing loomed—the military conducted ninety-eight tests.) Two hundred and ten of the explosions were aboveground, eight hundred and thirty-six were underground, and five were underwater, as the military experimented with bigger bombs, better designs, and different delivery mechanisms. The nukes blew up in the Pacific, the south Atlantic, New Mexico, Alaska, Mississippi, Colorado, and Nevada, where a government site held a staggering 925 tests (one hundred of which were atmospheric). These tests were witnessed by some 220,000 official participants—to say nothing of nearby civilian populations.

Citing local cancer rates that are six to eight times the national average, the protesters outside my car want New Mexicans to be included in the 1990 Radiation Exposure Compensation Act, which provides assistance to nuclear-test participants, uranium workers, and those downwind from the Nevada Test Site. A 2009 study by the

Centers for Disease Control found that the Trinity test exposed parts of New Mexico to ten thousand times the radiation permissible today. The CDC has also reported that everyone born after 1951 in the continental U.S. has received radiation—in every organ and tissue—from now-banned nuclear tests, the residual fallout from which will eventually kill some eleven thousand Americans.

After the test, Groves sent word to Washington, which cabled the preliminary news to Potsdam, where Truman was meeting with allies Stalin and Churchill to discuss the end of the war. (The conference had already been delayed once in the hopes of having the bomb completed.) The next day, Washington provided further details using a real-estate code: "Doctor Groves has just returned most enthusiastic and confident that the little boy is as husky as his big brother. The light in his eyes discernible from here to Highhold and I could hear his screams from here to my farm."

By the day after the test, fifty-seven Manhattan Project scientists had signed a petition asking Truman to consider the morality of using such a weapon. As the bomb was prepared for Japan, someone asked Oppenheimer why he was so glum; he replied, "I just keep thinking about all those poor little people." Some scientists were told the bomb would spare hundreds of thousands of U.S. soldiers that could be lost in a ground invasion; they did not sign the petition. Nor did physicist Edward Teller, who had decided, "The things we are working on are so terrible that no amount of protesting or fiddling with politics will save our souls."

The petition was circumvented by General Groves, who ensured it took a circuitous route to Washington, where the president wasn't in residence, anyway.

The remnants of Jumbo rest in the parking lot just outside the test site. Once weighing 214 tons of banded steel, Jumbo was a $12 million, twenty-five-foot-long burrito-shaped container that originally was meant to house the Gadget—the idea being that, in the case of a fizzle, the precious plutonium would be saved. Fabricated in Chicago, Jumbo was the heaviest single object ever shipped by rail. The top-secret, canvas-covered behemoth wound its way on a flatcar from Ohio to Louisiana to Texas to New Mexico, where it was tugged across the desert on a custom sixty-four-wheeled trailer. By the time it arrived a few months before the test, Jumbo was deemed unnecessary. Eventually, it was stood on one end in a tower eight hundred yards from Zero. The structure was vaporized, but Jumbo remained. In fact, the army has had a hard time getting rid of the shell. An attempt to explode it in 1947 only popped the ends off; Jumbo was buried, dug up, and eventually dragged to the parking lot in 1979, where it now sits on its side. I walk through it—a tunnel with fourteen-inch-thick walls—and think of close calls. Had Jumbo been used, it would have become another 214 tons of fallout.

The soldiers sell brats and burgers, shirts and shot glasses. They are young and polite, calling everyone "sir" and "ma'am." A record 5,534 visitors will show up today—even more than came twenty years ago for the fiftieth anniversary. (Later, I will shake hands with Brigadier General Timothy R. Coffin, the commander of the missile range, who stands tall in digital desert camo with his name—Coffin!—on his chest as he tells me each open house costs the army $60,000.) Meanwhile, people take selfies in front of radiation warnings posted on the chain-link fence. A film crew walks by, speaking Japanese. A woman's ball cap reads WE NEED MORE HEROES beneath an American flag.

Three weeks after the Trinity test, on August 6, 1945, at 8:16 a.m., the Little Boy uranium bomb exploded above Hiroshima. Three days later, the Fat Man plutonium device was dropped on Nagasaki. Estimates vary, but by 1950 some 200,000 were dead at Hiroshima and 140,000 dead at Nagasaki. The latter blast was bigger—at twenty-two kilotons—but the damage was mitigated by the surrounding hills.

Truman was at lunch when he received news of the first successful atomic bombing; he told his tablemates, "This is the greatest thing in history." In a radio address, he would explain to the nation, "The force from which the sun draws its power has been loosed against those who brought war to the Far East. . . . We have spent more than two billion dollars on the greatest scientific gamble in history—and we have won. But the greatest marvel is not the size of the enterprise, its secrecy, nor its cost, but the achievement of scientific brains in making it work."

After the war, the U.S. Strategic Bombing Survey decided that dropping the bomb had not been necessary to defeat Japan. The country would have surrendered on its own most likely before November—even without any plans for an Allied invasion. (The firebombing of Tokyo earlier in the year had killed more people—and destroyed more square miles of city—than the immediate effects of either atomic blast.) Before Hiroshima, General Eisenhower told Secretary of War Henry Stimson that nuking Japan was "completely unnecessary" and "no longer mandatory as a measure to save American lives." Many military voices echoed his thoughts. In a press conference six weeks after Nagasaki, General Curtis LeMay, the man in charge of the bombers, insisted that "the atomic bomb had nothing to do with the end of the war at all."

One British physicist would remember General Groves saying—in 1944—that the bomb's "real purpose" was "to subdue the Soviets," our allies in the war. The following year, about a month before he became secretary of state (and more than two months before Hiroshima), James Byrnes explained to a Manhattan Project scientist that using the bomb would make "Russia more manageable in Europe." In 1949, P. M. S. Blackett, a British Nobel Laureate and wartime advisor, wrote, "The dropping of the atomic bombs was not so much the last military act of the Second World War as the first major operation of the cold diplomatic war with Russia now in progress." In August of that year,

Russia tested its own atomic bomb, dubbed "First Lightning," which fused the soil of the Kazakh steppes a startling blue-black.

Some say Truman didn't really make a decision—that the bomb was so costly that it had to be used. He never doubted his actions and in fact wrote in 1963, "I would do it again." In late 1945, when Oppenheimer told the president, "I feel I have blood on my hands," Truman cut off dealings with him, calling him a "crybaby."

For an atomic bomb, you can either split the nucleus of a heavy atom like uranium or plutonium through fission, or you can smash together two lighter nuclei, such as hydrogen, to form a larger nucleus. This process, called fusion, powers the stars—and, thanks to Edward Teller, today's thermonuclear bombs. Many modern nuclear-weapon designers have backgrounds in astrophysics.

From the beginning, Teller was always more interested in the theoretically far more powerful fusion bomb, which he dubbed the "Super"—to the point that he dragged his heels on the Gadget. Fusion weapons are sometimes called hydrogen or thermonuclear bombs. The basic design—as envisioned by Edward Teller and Stanislaw Ulam—has leaked over the decades. A thermonuclear weapon is a staged device: a standard fission bomb implodes, the radiation from which compresses a secondary stage, which also undergoes fission—producing such temperatures that the hydrogen fuel caught between the two reactions must fuse. In other words, the legacy of Trinity—the implosion fission bomb—remains at the heart of every nuclear weapon.

After the Trinity test, Teller would ask Oppenheimer to support his quest for the thermonuclear bomb. Oppenheimer refused. In 1954, when the eccentric and left-leaning Oppenheimer got caught up in McCarthy-era security hearings, Teller would testify against him, leading to Oppenheimer's clearance being revoked. The scientific community was scandalized. Wernher von Braun, the father of rocket science, would tell Congress: "In England, Oppenheimer would have been knighted" for his service to his country. The atomic cowboy would never again work for the government. He died of throat cancer in 1967.

During the war—while he and his colleagues worked on the Gadget—Teller would brood over bigger and bigger bombs. On a blackboard, he tracked his ideas for thermonuclear weapons, including one known simply as "the Backyard," a planet-killer so destructive that it wouldn't even have to be launched against one's enemies, but instead could simply be detonated at home.

The U.S. first tested a full-scale thermonuclear device on an atoll in the Marshall Islands on November 1, 1952. The fireball alone from the 10.4-megaton blast stretched for three miles, while the mushroom cloud spread a hundred miles wide. The fallout from a thermonuclear bomb is believed to be even more lethal than the blast. (Many Marshall Islanders would die before being evacuated; thousands remain in exile today.) Nine months later, the Russians tested their first thermonuclear weapon, known as "the *Sloika*," or layer cake. The force of these bombs can also be boosted by injecting the cores with tritium and deuterium, and by using casings that also undergo fission. The largest thermonuke ever tested was fifty megatons—or twenty-five hundred times the force of Trinity. Nobel Laureate I. I. Rabi once said, "The world would be better without an Edward Teller."

Last summer, in the bookshelves in the den of my father's childhood home in Gettysburg, I found a copy of the paperback *How to Survive an Atomic Bomb*, published in 1950, the cover showing a nuclear family (a term that predates Trinity: here, a father, mother, sister, brother) standing united to bravely face, if not a new shining day, then perhaps the initial blast, the front of their bodies lit by a searing light (in which case, they're most likely goners). Written by a naval Senior Radiological Safety Monitor, the book promises no "scare talk" or rumors. It claims to deliver facts. ("The truth is bad enough—but nowhere near as bad as you probably think.") The cover proclaims, "If there's ATOMIC WARFARE this book may save your life!"

How to Survive an Atomic Bomb is written as an imaginary dialogue, a kind of nuclear catechism. ("What is the heat of the bomb like?" "Is it all right to smoke?") The expert might patronize ("It's pronounced 'ee-VAK-u-ate'"), bully ("Use your head"), or be ludicrously unhelpful. (What to think about while you're on the ground waiting for the blast? Try reciting jingles, multiplication tables, or—best of all—the steps to follow after an attack.) Radiation is like liquor: "A little bit won't do you any harm, but a lot of it will." The book covers three kinds of atomic clouds, the effects on livestock (because of their coloring, white leghorn chickens fare better than Rhode Island reds), and the results of certain doses of radiation. ("One morning you might look at your pillow and find that your hair had begun to fall out. . . . You'd also run a fever, and your bowels would run, and you'd feel rotten and 'achey' all over. You might even have bloody spots on your skin and slight bleedings in your mouth. . . . You might find that for a time you were unable to beget children, although you could still have sexual relations. *All these troubles would go away in time.*") There are chapters for those living in apartments, houses, and the country. Stark illustrations show fools falling off a fire escape, a farmer lying in a furrow, and how a fedora might save half of a man's face from the flash ("In time of war, if you work outdoors you should always be fully clothed"). There are several mistruths: "There is one fact you must remember—and it definitely is a *fact.* Not one person in Hiroshima or Nagasaki was killed or injured by lingering radioactivity." Also: "Facts will help kill the fear that causes panic."

I join the quarter-mile march to the fenced oval that is Ground Zero. An Asian woman drags a wailing little girl by the arm. The crowd circles the lava-rock obelisk, erected twenty years after the fact on the spot where the Gadget's tower stood. A guy in a NASA shirt holds a yellow Geiger counter. (A brochure I received upon entering the range reminds me that "although radiation levels at Ground Zero are low"—no more than ten times the natural background amount—"some feel any extra exposure should be avoided.") Trinitite is still easy to find. A sign threatens thieves with fines and jail. Some teens huddle around a large green pile, which they have collected and washed off with water. I spot a fair number of little kids, plus at least three or four babies. Ahead of me, a father struggles to push an infant in a stroller through the sand, from which europium, cesium, cobalt, strontium, and plutonium are currently emitting alpha, beta, and gamma radiation. (The last rays are stopped only by one inch of lead or eight of concrete.) Historical photos hang on the far fence. A bomb casing similar to Fat Man's sits on a trailer. Later, I will ride a hot, dusty bus two miles to the ranch house where the scientists assembled the core. There, a guy will look to the mountains and say, "I figure it would make you goofy, being out here." Inside the house, I'll hear a voice from another room offer some kind of summary: "It's just the power of nature, the power of God, whatever you want to call it."

There is no doubt atomic blasts are aesthetically beautiful: incomparable and illicit expressions of nature's hidden physics—the micro blown up to such a macro scale that something, or maybe everything, inside you is stirred. In time, awe gives way to a grim connoisseurship. As it is so often said, fear quickly becomes something like desire. I could watch the declassified films for hours. There are so many stunning details: how the air rushes forward—with the blast wave—then blows briefly backward as it is sucked up to form the towering mushroom. That's how I feel watching these awful films: pushed and pulled—shocked, devastated, repulsed, but then drawn into the cloud.

I stumble across an amazing artifact from the National Archives: a black-and-white army film that contains the actual sound of the blast. I put on headphones and I'm at Yucca Flat, Nevada, in the early morning of March 17, 1953, for the *Annie* test of Operation Upshot–Knothole, which was open to reporters, who watched 7.5 miles away from "News Nob." The announcer counts down to the flash, which is blinding. Thirty seconds of murmuring as the sixteen-kiloton fireball climbs. *Oh, look at that. Oh boy, George.* The cloud billows up. *Woo-hoo!* The announcer warns the shockwave will soon arrive. Suddenly, a shotgun blast that crescendos louder and louder to become a rolling thunder. When it has receded, a man shouts, "Holy shit!" Then he shouts it again.

There were tests in the desert and under the sea. Rows of burning trees whip one way, then the next. Plumes of roiling water toss battleships like toys in a tub. In an instant, the paint on a bus burns off before the smoking black frame is flattened by the blast. Heat sears rows of caged pigs wearing military uniforms, the exposed flesh meant to mimic human skin. (Imagine some twelve hundred porcine subjects in 1957 alone. Sheep and monkeys, too, the latter's eyes taped wide open.) From inside a house, there's a flash at the window; the blinds billow gently inward, then begin to smolder—before the entire structure is blown away. More interior scenes with nattily dressed mannequins posed in quiet domestic dramas: couples having cocktails, Junior riding the arm of the couch, Baby penned in front of the TV, Mom sitting dreamily at the window, moments before the panes impale her and slam her body across the room.

Supposedly named after everything from ghost towns to cheeses to prostitutes, a litany of tests runs through my head—Castle Bravo, Ivy Mike, Dormouse Prime, Little Feller, Diamond Fortune, Buster/Jangle, Tumbler-Snapper, Teapot, Wigwam, Redwing, Plumbbob, Romeo, Nougat, Gnome, Zucchini, Muenster, Diablo, Shasta, and Sugar—countless atomic suns born across the land. After a while, I can tell the clouds apart. Air tests, tower tests, underground tests—plus the atomic cannon (fired once). Soldiers would wait in trenches until the shockwave passed, then march into the blast zone and perform maneuvers beneath the fallout, while paratroopers floated down.

I watch the films; I sift through photographs. The images veer from the inconceivable to the insane. A live fish so radioactive that it made its own X-ray (the algae it had recently ingested glowing in its stomach). Marines clowning around for the camera, their hands "holding up" a mushroom cloud. Rows of VIPs in goggles and khaki shorts, watching from Adirondack chairs, their stern faces lit by the flash. A ballerina in a black leotard performing a *pax de deux* with the atomic cloud that climbs over her shoulder. Aerial views of the giant collapsed craters that pock the desert—like dimples on the moon—at what used to be known as the Nevada Test Site, north of Las Vegas. (The holes, carved out of the earth by more than eight hundred underground blasts, are still visited fondly by the aging scientists who made them.)

The height of bravado: a 1957 film of five volunteers standing in a huddle while a F-89 jet fires a two-kiloton nuclear air-to-air rocket that explodes 18,500 feet directly above them. A handmade sign stuck in the dirt reads GROUND ZERO. POPULATION: 5. The men shield their eyes with their hands; only one bothers to wear sunglasses. The center man delivers the play-by-play. After the blast, the men rejoice with hearty handshakes; one produces a cigar. The honey-voiced announcer calls the experience "just a wonderful thrill," while the colonel effuses for those watching at home: "My only regrets right now are . . . that everybody couldn't have been out here at Ground Zero with us." All five men would eventually contract cancer.

The U.S. even detonated at least six nukes in space, shorting out terrestrial electronics and creating dazzling atomic auroras that played across the globe. In 1963, the Limited Test Ban Treaty banned nuclear testing aboveground, underwater, and in space. In 1992, the U.S. conducted its final underground test, though the Comprehensive Nuclear Test Ban Treaty—signed by President Clinton in 1996, but blocked by the Senate—still awaits ratification.

Maybe we all go a little mad in the desert of our imaginations. One morning, as the test date neared, Oppenheimer found everyone outside, staring at an unknown object blazing in the sky. The men raced for binoculars. A spy craft? Sabotage? From Albuquerque, Kirtland Air Force Base reported they had no planes able to intercept it. Oppenheimer recalled, "Our director of personnel was an astronomer and a man of some human wisdom; and he finally came to my office and asked whether we would stop trying to shoot down Venus."

Later, in 1960, NORAD went on highest alert—99.9 percent sure of a nuclear attack—when a U.S. Air Force radar station in Greenland mistook the rising moon for a Siberian missile launch.

An arms race is a search for military high ground. *A Study of Lunar Research Flights, Volume 1,* issued from Kirtland Air Force Base on June 19, 1959, outlines a secret Air Force plan to detonate a nuke on the moon. The idea—officially known as Project A-119—was to offer a show of American strength in the face of Sputnik. ("Obviously . . . specific positive effects would accrue to the nation first performing such a feat.") A young Carl Sagan contributed to the study, which outlines the scientific, military, and political gains in hundreds of pages of equations, measurements, tables, and arguments, all contextualizing the central madness: launching a nuclear missile at the moon. The report covers potential hazards in appendices such as "Survival Time of an Irradiated Population" and "Current Attitudes and Activities Regarding Biological Contamination of Extraterrestrial Bodies." Volume two of the report remains classified, while other studies have been destroyed.

Thankfully, instead of nuking the moon, the U.S. decided to try to walk on it, which we did in 1969, as the Cold War fueled the American race to space. Two days before my visit to Trinity, at an Albuquerque restaurant, I met the last man to step onto the moon, Apollo 17 astronaut Harrison H. Schmitt, who spent more than three days in 1972 on the lunar surface. I asked Schmitt if he was familiar with Project A-119. "I hadn't heard of that one," he said, laughing. "Ah, just another crater." Then he reminded me of another atomic detonation in New Mexico, an unsuccessful attempt in 1967 to use a nuclear bomb to frack a gas well in the northwest corner of the state. He laughed again. "Nukes don't frack; they melt."

Schmitt was ten when the Trinity bomb was tested. His family lived outside of Silver City, some 150 miles away, where he woke up to not one but *two* blasts that morning. (The second was a reflection—another echo.) The detonation occurred during a shift change at the mine where his father worked as a geologist, and Schmitt remembers him coming home for lunch and scoffing, "I don't know what that was, but that was no ammo dump." After the war, Schmitt would watch the V-2 contrails rise over White Sands. He would grow up to meet Wernher von Braun, Robert Goddard, and other space pioneers before eventually riding a Saturn V rocket from Florida to the moon.

In 1982, as a U.S. senator, Schmitt would help broker the settlement between the government and the David McDonald family, the last of nearly one hundred families whose land had been seized to make the Trinity test site and bombing range. The land was never returned to the ranchers, who had been undercompensated. In protest, David, eighty-one, and his niece, Mary, reoccupied their ranch—which included the old site of the Trinity base camp. Schmitt intervened in the armed standoff, explaining, "I just tried to encourage people to talk to each other."

Over lunch, the spaceman and I discussed science in the desert—from the ancient Anasazi stargazers in Chaco Canyon to the Very Large Array radio telescope in the plains west of Socorro—and whether the U.S. has lost its will to explore. A geologist like his father, Schmitt is the only professional scientist to have landed on the moon. In the lunar rover, he explored nearly nineteen miles of the Taurus–Littrow valley, where he and his commander collected 243 pounds of material. Despite the grueling demands of the mission, Schmitt's boyish enthusiasm shone through, just a giddy scientist at play. NASA videos show him running, bouncing, and skiing down hills, singing goofy songs ("I was strolling on the moon one day, in the merry, merry month of May—no, December!"), and begging Mission Control to let him hurl his geology hammer in the moon's one-sixth gravity. Over lunch, he waxed on about the valley's deep beauty, the slopes brilliantly illuminated against the black sky, where the earth always hung in the same position. At one point on the mission tapes Schmitt deadpans, "You seen one earth, you've seen 'em all." He took the famous "Blue Marble" photograph, an image that is often credited with raising global consciousness and underscoring the fragility of our home planet. The horizontal tread of his footprints will stand on the moon for millions of years.

Most of all, Schmitt remains frustrated that we haven't been back. His book, *Return to the Moon,* advocates the development of a program to collect lunar helium-3 as a source for fusion power plants on earth. He believes all interplanetary roads—settlements in space, tourist flights to Mars—lead through the moon. ("The moon's the place to learn how to do it.") He talked about what it would take for the U.S. to maintain its space dominance: a younger, nimbler space agency, for starters. ("Apollo was a young person's program—the average age was around twenty-five.") That said, he no longer thinks the government is the best driver of exploration. He favors private funding, though he's wary of his state's massive spaceport, telling me at one point, "They have stars in their eyes."

What do you ask someone who has left this world behind? I can only imagine what Schmitt thinks when he looks at the night sky, or what he felt leaving the moon, knowing he would never be back. ("All of us knew we would get one shot, one mission. But it was beyond our imagination that the whole technology would be abandoned.") During our lunch, he told me something startling. He said the reason scientists—including ones at my own university—continue to pore over samples he brought back forty years ago is that looking at the moon is a way of peering into our own vanished past. He explained, "The moon is the history of the earth that we can no longer see." The moon is constant, its every experience written on its face. And what does the record reveal—what does creation look like? Schmitt raised his eyebrows: "It was *extraordinarily violent.*"

A man in a denim jacket describes molten Trinitite raining down and collecting in puddles on the ground. His name is Robert Hermes, and he's a retired Los Alamos physicist. He has several chunks of Trinitite set out on a table before him, plus a sizable piece pinned to his hat. He holds up a tube of green beads he found in the ant sand this morning. (Anthills are notoriously radioactive, as the insects pile up Trinitite.) He says such glassy droplets appear worldwide after other big impacts, particularly extinction events. He rattles the tube. "They're the same size as the beads they found off the Yucatán—where the asteroid hit that killed all the dinosaurs."

What is it about vast, empty spaces that beckons the imagination? Why do we choose to fill them with the seeds of our destruction? The desert. The moon. Twin two-sided coins.

I grew up in a city, but I spent a few summers at a camp in New Mexico, and as a teen I camped in the Pecos Wilderness, not far from Los Alamos. I remember never having seen so many stars; they stretched over my head like a blanket—a sky composed of more light than darkness. Or so it seemed then. At that moment, I had a profound realization of what I had been missing—and how much of the universe I would never live to see. It was an idea at once sad and thrilling.

Now my daughter dreams of going into space (along with being a superhero and catching a fairy). She wants her mother and me to come with her, of course, because she's only four. So we go to the Science Center and sit in rocket chairs and watch the monitors as we "blast off" into space. They say someone her age might grow up to walk on Mars.

The U.S. occupational radiation threshold (for those working with nuclear power, medicine, weapons) is more than double the international limit. Some critics want it raised even higher, arguing the damage done by radiation is not linear—that there is a level at which long-term exposure is safe. (These skeptics, such as Representative Lamar Smith, a Texas Republican who chairs the House Science, Space, and Technology Committee, would loosen regulations on nuclear industries.) Meanwhile, scientists have discovered a "bystander effect": irradiate a single cell, and those around it will suffer. In other words, cells communicate. The damage spreads wider and wider, every one a casualty.

In total, I will spend an hour within the inner fence at Ground Zero, which means I will absorb somewhere between one half to one millirem—at most, a sixth of a chest X-ray, or the cosmic radiation one receives flying the red-eye across the country, or what we soak up in a year sitting in front of the TV. A webpage from the Nuclear Regulatory Commission will tell me I have shorted my life expectancy by a minute and twelve seconds. On my way out, I will bite down and feel grit in my teeth. I'll try not to lick my dusty lips, or think about the half-life of plutonium. I'll blow my nose, rinse my mouth, and spit once I'm back at my car. But long after I've showered and scrubbed and left Trinity behind, I will feel its dust in my nose and taste its dirt in my mouth.

I'm wary of the line I'm treading. I don't want to fall too hard for the atomic mystique, or slip into radiophobia—romanticizing the danger of an often-misunderstood science. Nobody would wish all radioactive materials from the earth. Radiation warms the planet from without and within. Many things are naturally radioactive, from bananas to our own bodies. And I wouldn't want to live without nuclear medicine—its imaging, scans, and treatments. I stood next to my daughter when she got her first X-rays—for a mysterious pain in her leg that was temporarily crippling but ultimately benign. Only a toddler, she asked, "Daddy, why don't I get to wear a big apron?" I held her hand while the huge machine made its ugly noise over her tiny bare body, and I remember being grateful for pictures of the unseen.

Back inside the fence, a retired White Sands public affairs officer is sitting—in some official capacity, I imagine—on a stool from which hangs a sign promising FREE ANSWERS. A freckled girl wrapped in a blanket runs a green rock up to him. She smiles when he says, "Yes, that's the stuff, all right." Suddenly, a Japanese interviewer puts a microphone in his face. A cameraman is rolling. We all fall silent. The interviewer doesn't smile, but asks, "What is purpose of opening the site? What do you want the people to know?" The answer man thinks for a moment, then says, "We want people to understand what happened, and why it happened, and then go away to make their own judgments."

During the war, the average age at Los Alamos was twenty-seven. Over the years, many of the scientists would be gripped by nostalgia for the days when they were young and working on the bomb. Edward Teller would say, "In spite of the difficulties, I (and many others) consider the wartime years at Los Alamos the most wonderful time in our lives." A British physicist recalled, "Here at Los Alamos, I found a spirit of Athens, of Plato, of an ideal Republic." Nobel Laureate Hans Bethe said at Oppenheimer's memorial service, "There were other wartime laboratories of high achievement. . . . But I have never observed in any one of these other groups quite the spirit of belonging together, quite the urge to reminisce about the days of the laboratory, quite the feeling that this was really the great time of their lives." As one physicist remembered in 1970, "It was one of the few times in my life when I felt truly alive."

Lately, a similar Cold War nostalgia has slipped into the national rhetoric, as politicians and pundits yearn for the less confusing world when it was just us versus them. How quickly we forget—and confuse simplicity with safety. I grew up a Cold War kid. In the 1980s, the threat of nuclear war was always in the background, like the TV. Among my childhood possessions, I recently found a 160-page booklet issued by the Department of Defense called *Soviet Military Power.* Packed with graphs, stats, war plans, and lavish illustrations of tanks, bombers, missiles, subs, and satellites, the book lingers over our enemy's might with a lover's eye. Today, I find the book unsettling. I picked up *Soviet Military Power* in 1987—the year it was published—when I was ten.

The day grows long. We have been reduced to a bunch of tourists milling about under military gaze. A small crowd stands around the obelisk, which is less of a focal point than a photo op. *(We considered ourselves to be a powerful culture.)* People approach warily, pose, then retreat. Nobody takes turns; everyone goes home with pictures of each other, which is as it should be. We're all in this together, a kind of bystander effect. *(This place is not a place of honor.)* Beside the monument stands a fading footprint of concrete and rebar—the remains of a leg of the tower. I don't think of the invisible rays that may or may not be killing us. *(What is here is dangerous and repulsive to us.)* We're all dying anyway, one way or another, as the great clock counts down. *(The danger is still present, in your time, as it was in ours.)*

What is the half-life of memory? The observation bunkers, where Oppenheimer and the scientists fretted then rejoiced, are long gone. Creosote, yucca, and yellow grass are reclaiming the plain. At Ground Zero, the hot dust decays, leeching energy into the desert. The event moves into the mind, but we must hold onto the dead: Hiroshima. Nagasaki. The scientists and soldiers who tested and were tested upon. The rest of us standing downwind of history, ongoing casualties of a never-ending Cold War. We've come up with so many clever ways to wipe each other off the earth.

Standing here gives only a feeling of emptiness. Not necessarily despair, or sorrow, but absence. We've all come to Zero. No thoughts, no words—nothing remains. Complete devastation. And isn't that the point of the bomb?

2015

Chasing the Boundary: Boom and Bust on the High Prairie

Once again, the story begins with fire, some primeval attraction to light in the darkness. A time-lapse video looking down at the nighttime earth, courtesy of astronauts in the International Space Station spinning some 250 miles above. Civilization stretched out like campfires, cities shining in the dark, great webs of light marking our dubious progress across the dim globe: the golden Nile dangling down Egypt, glittering island flecks in the Philippine Sea, Italy and its boot, the Holy Land glowing against the black Mediterranean, the floodlit border between India and Pakistan. Meanwhile, unlit places hold their vast dignity: the Amazon basin, the steppes of Eurasia, the deserts of Africa, the south Indian Ocean. Green auroras play across the thermosphere. Lightning sparks the clouds. Across the United States roll the wide fields of the republic: the dazzling eastern seaboard, the dark of Appalachia, the twinkling townships gridding the plains. And there—to the west of the Great Lakes (blank in the night) and over the bright shoulder of Minneapolis–St. Paul—a strange bloom. Like a city but bigger, more diffuse. No central pinprick but countless tiny suns. A region smeared with light. Once you know what you're looking for, you can't miss it as the camera swoops over—but what is it? Nothing was there seven years ago. This is a new shining star of the north—the blazing rigs, equipment, settlements, and gas flares of the North Dakota Miracle, a.k.a. the Oil Patch, a.k.a. Kuwait on the Prairie, a.k.a. the Bakken Oil Boom, U.S.A.

The phenomenon had been covered in the news, by the networks, on the radio, and in the glossies: North Dakota was a new kind of promised land. This was no ordinary boom. In 2013, North Dakota's real gross domestic product grew 9.7 percent, easily beating the national average (1.8 percent). Meanwhile, the personal income of a North Dakotan rose 7.6 percent, the greatest increase in the country—a distinction the state has boasted for six of the past seven years. Per capita personal income was second highest in the nation, after Connecticut. This wasn't just a few fat cats getting richer—there were *jobs.* North Dakota had the lowest unemployment rate for the fifth year running, at 2.9 percent, and joblessness was even lower in the oil counties (at 1.6 percent). The national average was 7.4 percent. Taxes on oil and gas production had allowed the state to sock away $2 billion in savings.

And so the people came. In 2013, North Dakota was the fastest-growing state, increasing its population at more than four times the national average. Williams County, which includes the city of Williston, in the middle of the Bakken oil field, was the fastest-growing county in the country with a population of more than ten thousand. And while the country as a whole is aging, North Dakota is growing younger, thanks to an influx of twenty- and thirtysomethings migrating to the state. In fact, North Dakota is changing so quickly that the director of the U.S. Census Bureau made a special research trip in May, nine months after his confirmation, to wrap his head around the miracle. His finding: "The growth was truly like nothing I have ever seen."

And so for many the summer of 2014 was a time of brave promise. The boom had carried North Dakota through the last recession and beyond. As the state looked toward its 125th birthday—in November, though to be celebrated in August, because even North Dakotans don't want to party in *that* kind of cold—its populace was in the grip of a spirited flurry of civic-minded activity: games of birthday bingo, a wagon train, cowboy poetry, military encampments, living history, a "125th statehood anniversary shoot," appearances by a "nationally renowned" Teddy Roosevelt impersonator, roughrider days, rodeos, geocaching, quilting, a two-state book club (both North and South reading *Dakota: A Spiritual Geography*), and music of all kinds, including a newish track, "North Dakota"—*Now we're from a town you've probably never been / We're from the north, but we're not Canadian*—from the homegrown pop-country teen sister act, Tigirlily, described on their website as "optimism personified."

But not everyone was optimistic. Not everyone was cheering. The narrative had somehow gone awry. Did the boom offer promise or peril? Men were murdering women and rumored to be raping other men. Prostitutes and strippers flocked in from out of state. Cartels were moving meth and heroin. Citizens were killing themselves at an alarming rate. Trucks were crashing, trains exploding. People weren't just disappearing—they were vaporizing. A Facebook page called "Missing Persons & Property from the Bakken Oilfield" offered a litany of loss: dogs, trailers, mothers, minors, fugitives, tools, trucks, more dogs, equipment, ATVs, snowmobiles, and men, men, men.

So I went north in late July of that anniversary summer to see what was going on.

There isn't much to the Bismarck airport. A *Triceratops* skull sits in a glass case on the way to baggage claim, evidence of the state's rich fossil life. Down the escalator stands a banner ad for a pipe company founded in Dubai. The man ahead of me harasses the kid behind the rental counter about the price of gasoline: "But ya'll are pulling it out of the ground up here!"

The Bakken Formation is a layer of 360-million-year-old rock that stretches some twenty-five thousand square miles across the U.S. and Canada; it is the largest continuous oil accumulation assessed by the United States Geological Survey. About two-thirds of it lies below North Dakota, in the northwest corner of the state. Its discovery dates to the 1950s, but only very recently have rising oil prices and new technology—such as hydraulic fracturing and horizontal drilling—made extraction feasible. In 2006, North Dakota ranked ninth in domestic oil production. Now it's number two, behind Texas. Republican Governor Jack Dalrymple has compared his state to a "small OPEC nation"—and, indeed, it now pumps a million barrels a day, more than OPEC members Qatar and Ecuador.

Within minutes of leaving the airport, I cross the Missouri River. The speed limit soars to seventy-five miles per hour as I skim across the prairie. Grasslands rise and fall to the horizon, a land without trees. Clouds cruise overhead like battleships, casting shadows on the highway. The wind buffets the hills, bending the grass in waves, a choppy green sea. Thanks to a wet spring, the wheat crop is about two weeks behind, but a cool summer has the farmers hopeful. Hay bales dot the fields, great strips of the prairie rolled up. The grass takes on many shades—neon, avocado, mint, olive, shamrock. Unbeknownst to me, I have been living back home in a monochrome of green. I see an oil train to the south. The radio plays an ad—truck drivers are needed.

Here's another way to understand the pace of the boom: In 1951, North Dakota got into the oil business. It took thirty-eight years to produce a billion barrels, and twenty-two years to barrel a billion more. Its third billionth barrel, which arrives in 2015, will only have taken four years. Crude oil production quintupled between 2007 and 2012.

Currently, there are more than ten thousand wells operating in North Dakota—a number that could quadruple in the coming decades. The month I visited, more than 190 rigs were actively drilling new wells. A Bakken well requires more than two thousand truckloads of equipment and material in its first year, after which oil and water need to be hauled—in and out—for the life of the well. The state's infrastructure is unprepared for such traffic: highways are cracking, potholes and litter abound, and dirt and gravel roads are kicking up so much dust they're smothering crops and animals. (Ranchers have reported cattle dying of dust pneumonia.) Since 2008, crash fatalities have more than doubled in oil counties, while severe injuries from truck accidents are up 1,200 percent.

In many parts of the state, drivers "wave" at each other—just two fingers raised off the wheel, less a gesture of overt friendliness, I think, than some sort of existential affirmation ("I see you, you see me—we both exist in this vast, empty landscape that threatens to argue otherwise"). But this gesture dies as you drive north into the Bakken, where trucks are many and time is money. I've never seen such industrial traffic: trucks hauling sand, storage tanks, pipes, tractors, earthmovers, even other trucks—countless "oversized loads." Signs at the edge of the highway point down dirt roads to the oil rigs. Tanker trucks whip up clouds of dust so thick it looks like smoke. My rented Corolla shudders when the tankers pass. Going down a hill, I'm sandwiched between two semis hauling enormous trailers jacked up at what looks like a perilous degree. One in front, one behind—they start to squeeze me in.

The Bakken is a tight oil play: roughly two layers of black shale with a silty band of sandstone/dolomite sandwiched between, where the shale oil collects. (The geologist who discovered and named the Bakken, in the 1950s, compared it to an Oreo cookie.) The formation is deep (about two miles belowground), thin (only some 140 feet thick, with an even thinner middle layer), and relatively impermeable, meaning the crude does not flow easily. That said, last year the United States Geological Survey doubled its estimate of the total amount of undiscovered, recoverable oil in the Bakken and the Three Forks Formation just below it to 7.4 billion barrels (as well as 6.7 trillion cubic feet of natural gas). This year, a more bullish oil executive put the estimate of recoverable oil at more than four times that amount.

South of Williston stands a twenty-foot-tall head of Abraham Lincoln plopped down in an RV park by a developer from Brooklyn, who turned his back on Big Apple real estate in favor of the Bakken, where he told a reporter last year he was making a 100 percent return on his investments. In 2013, the number of housing units in North Dakota grew the fastest in the country at a rate of 3.1 percent. But it's simply not enough—the influx of people has fast outstripped available housing. In three years, the population of Williston has grown 41 percent to 20,850—though some unofficial estimates put it at nearly three times that. At the beginning of 2014, Williston boasted the highest average rent *in the country* for an entry-level, one bedroom, seven-hundred-square-foot apartment, which went for $2,394 a month—or nearly $900 more than the same space in New York City. Hotels, which are being built all the time, fill up. Since the boom, people have been living in campers, cars, and trucks—even through the impossible winters. Small-time "slum lords" rent out any open space to RVs and campers, despite substandard septic and electrical service. Some workers choose to commute hundreds of miles: two weeks on and one week off—when they fly or drive home. Others live in company housing known as "man camps," basically small cities of modular barracks—rows and rows of identical prefab trailers with plenty of food and flat screens, but no guests, guns, or alcohol. I pass three man camps in a row on the way to my hotel. The nearest camp, built in seventy-five days for workers from companies such as Halliburton, offers 415 beds in one of two floor plans: a "VIP/Executive Room" and a "Jack and Jill" model (with a shared bathroom)—though here it's all Jacks and no Jills. There are some nine thousand such temporary units in Williston alone.

The sign says WELCOME TO WILLISTON, N.D., BOOMTOWN, U.S.A. I am staying a few miles north of town at the Dakota Landing, a residential hotel that opened less than a year ago and caters to oil workers. (Its tagline: "Where tough guys sleep like babies.") It offers 240 rooms, twenty-four-hour food, a cafeteria, a bar/lounge, pool and ping-pong tables, business and fitness centers, block heater hookups in the parking lot, an indoor boot room (with lockers), and packed lunches to go. It's costing me about $175 a night after taxes—not bad for this town, although given that the average occupant is more of a resident than an overnighter, it's hardly a bargain at more than $3,800 a month (at the weekly rate). Across from the parking lot, an oil rig is going up.

The idea is relatively simple and ingenious, from an engineering standpoint. First, drill a well. As the well is dug, the wellbore is encased in cement to secure the pipe and prevent the seepage of fluid (into, say, local groundwater). Once the target depth is reached, the well is "completed." Methods vary. In some wells, a "perforating gun" is lowered, firing off shaped charges that blast small holes through the pipe and into the layer of rock. Other wells drop plastic balls down the shaft that open mechanical sliding sleeves in the pre-perforated liner, revealing strategically spaced holes. Whatever the case, the target rock can now be hydraulically shattered. A mix of water, sand, and chemicals is injected at such high pressure that it forces open fissures in the rock. When the liquid is removed, the particles of sand left behind keep the fissures open, allowing oil and gas to enter the well. Working backward, multiple sections of pipe can be "fracked," with a temporary plug closed after each stage is completed. When the last section is finished, the plugs are removed and oil and gas can flow to the surface.

Fracking is particularly effective when it is coupled with horizontal drilling, which allows you to drill down and then sideways through a single rock layer. This horizontal tail—often up to two miles long—can then be fracked in multiple sections, vastly increasing exposure to the oil-rich rock. Think: instead of dipping a chip into a bowl of seven-layer dip from above to get at the guacamole buried below, what if you could dip your chip sideways beneath the surface, swiping through the entire layer of rich green stuff?

There are Slavic accents behind the counter at the Dairy Queen, where my combo meal costs a surprising eleven dollars. Sticky-faced kids run around eating ice cream. In the park across the street, skate punks with ripped shirts, dreads, and dyed hair roll up and down a ramp in the golden early-evening light. Some of the younger ones enter the restaurant. Once they've got their cones, they go outside and stand next to some high school guys who are smoking. When the older kids aren't looking, the youngsters study how to exhale. A trio of Latinos in blue overalls lingers over dinner. The line for the drive-through stretches across the parking lot. A guy in jeans and dusty boots walks by, a gallon jug of water in each hand. A kid on a bike pops a wheelie. A baby kicks his high chair while his grandmother spoons him soft-serve. A beautiful summer Saturday night in a town on the American plains.

There have been heinous crimes. Last December, investigators say a Bakken oil speculator ordered a successful hit on a Washington state businessman who owed him two million dollars. (The triggerman was caught with a to-do list that read "wipe tools down" and "practice with pistol.") In 2012, two Colorado men on crack abducted a beloved Montana schoolteacher out for a jog, strangled her, bought a shovel at the Williston Walmart, buried the body in a shallow grave, and then returned the shovel. An extensive, well-armed, Williston-based meth ring calling itself "the Family" was broken up after it tried to beat one of its members to death. On the nearby Fort Berthold reservation, a sixty-four-year-old woman and her three grandchildren were killed by an intruder who—high on meth—slit his throat in front of the police (a fourth grandchild survived the attack by playing dead).

Meanwhile, down in Dickinson, a twenty-four-year-old Idaho man looking for work broke in and raped an eighty-three-year-old woman in her home.

Locals decry the influx of migrant workers, calling them "oil field trash" and "rig pigs." In the past five years, federal prosecutions have nearly tripled in western North Dakota. The FBI opened an office in the Bakken, which the DEA has identified as a "federal High Intensity Drug Trafficking Area." The ATF and Bureau of Indian Affairs have increased their presence, too. "Operation Pipe Cleaner." "Operation Winter's End." The arrests pile up. Between 2005 and 2011, calls to the Williston police department more than quadrupled; in nearby Watford City, they increased nearly one hundred times. In a North Dakota State University study released last year, one officer said, "I think one in every five people that I deal with has drugs on them." While I'm in Williston, a headline reads "DEA Watching North Dakota."

Admission to the Williston Basin Speedway is fifteen bucks, ten if you're military. You hear it before you see it—a steady roar punctuated with the percussive POP! of backfires. In the family area ("no alcohol or profanity allowed"), a weathered old woman with burgundy hair squints to see the track. Elsewhere, guys strut around in sunglasses even though it's 8:30 at night. Everyone's T-shirt announces something: a sports team, a car, a beer, or, in the case of one woman, that she's one of the "Bakken bitches." A guy in jean shorts wears a black tee with a MasterCard logo over which is printed "Sex with me: priceless." He walks hand in hand with a little girl in pigtails. On the track, two brothers from Minot are battling for first and second. The announcer plays hockey with them both: "They won't back down from a fight—even with each other." The concession booth sells a six-dollar "taco in a bag." Something fried stinks up the stands; they're burning the oil. Everyone is drinking silver buckets of iced blue cans of Bud Light. I'm not sure if one can even buy a single beer. Wedge-shaped cars with bright decals scramble and slide around in circles, throwing dirt. Over the PA, they play the Macarena. Some of us dance. The guy behind me—ribbed undershirt, neon shoes, blue ball cap, hand towel dangling from his back pocket—barks at his driver: "Come *on!*" The scoreboard counts down the laps. Faster and faster, going nowhere. Clods of dirt hit the fence. The noise is tremendous. Twenty-three machines pass us like peals of thunder. We feel it in the metal stands; we feel it in our guts. We shout at the cars, which deafly drive on. The roar crescendos and just when we think we can't bear it—it dies, just a bit, as the cars wing around the far side. We're on the last lap. The bouncing kid next to me has all of his fingers crossed. Down the row, a toddler's ears are covered with pink zebra-striped earmuffs; the baby's in the stroller next to her are not. We are choking on dust and diesel in this cathedral of the internal combustion engine and the fossils that fuel it. A checkered flag—the roar subsides. We clap politely for the winner, who came from behind at the last minute. Water trucks come out and spray down the track, a brief intermission. At the beginning of the next race, a driver spins out and crashes into the wall. Cars ram into each other, belching fumes. Two tow trucks race over. The announcer says, "I think he's done for tonight." The guy behind me blames the water trucks. He shouts, "A little too much water, guys. But, hey, we like it wet!"

Fracking requires great amounts of water—at least some two million gallons for the initial fracturing—that must be trucked to the well. Worse, Bakken wells typically require an additional six hundred gallons of freshwater a day to flush out salty precipitation and maintain the oil flow—which means a well could suck up another 6.6 to 8.8 million gallons in its three-to-four-decade lifespan.

North Dakota is one of the driest states in the country.

The "fracking fluid" itself is made of water, sand, and a proprietary mix of chemicals that companies are reluctant to disclose but usually contains a noxious brew of formaldehyde, benzene, acids, and volatile organic compounds. (The goal: kill microorganisms, prevent pipe corrosion, and increase fluid viscosity.) The majority of fracking fluid is absorbed deep underground, where it supposedly stays well below the water table. Ten to thirty percent returns to the surface.

Some of the surface "flowback" may be recycled (to frack more wells), but the majority of the wastewater—which now contains salty groundwater that can be naturally radioactive and contain heavy metals—must be hauled away, treated, diluted, or dumped into deep disposal wells. In the Bakken, bringing up a barrel of oil produces about a barrel and a half of briny wastewater. Wastewater is stored in open-air pits or tanks near the well—another potential source of groundwater contamination.

The idea of fracking isn't new, just improved upon. In the 1860s, Civil War veteran Colonel Edward A. L. Roberts patented a "petroleum torpedo" filled with black powder (later nitroglycerin) that—when exploded at the bottom of a well—vastly increased production. In the 1930s and '40s, other methods were developed to fracture the rock, including acid, steel bullets, and a "hydrafrac treatment" gel (a chemical mix of crude oil, solvents, and sand).

On December 10, 1967, as part of "Project Plowshare," the U.S. Atomic Energy Commission lowered a twenty-nine-kiloton nuclear device 4,240 feet down a natural gas well outside of Farmington, New Mexico. The device, named Gasbuggy, detonated with nearly twice the power of the Hiroshima bomb. It was the birth of nuclear fracking. Two years later, a forty-kiloton device was exploded down a gas well twice as deep in Rulison, Colorado. And in 1973, three bombs of thirty-three kilotons each were exploded nearly simultaneously at varying depths down a single well—like a string of firecrackers—outside of Rifle, Colorado. In each test, the nuclear blast vaporized the rock, creating a short-lived but, in my imagination, terrifyingly beautiful spherical cavern of molten glass some 150 feet in diameter that quickly collapsed into a tall rubble chimney. Nearly instantaneous boom and bust.

In all three tests, the target layer was widely fractured; greater amounts of gas were indeed "liberated," but the gas was radioactive—and unusable—every time. Today, the sites bear historical markers and, in some places, small parking areas at surface Ground Zero. The Office of Legacy Management monitors the land. The fact sheets for each blast state that the Department of Energy does not plan to remove the radioactive contamination deep underground because "no feasible technology" currently exists.

Even if the theory holds—and today's hazardous fracking fluids stay trapped below our drinking water—wells frequently fail for a number of reasons, including human error. Well casings—those barriers to seepage closer to the surface—often aren't cemented properly, which is part of what led to the blowout of the Deepwater Horizon. Studies suggest as many as 10 percent of all well casings leak.

Five minutes on the Facebook page "Bakken Oilfield Fail of the Day"—endless photos of trucks in the ditch, machinery upended, jerry-rigged equipment, workplace accidents, fires, and ridiculous wrecks—is enough to reduce anyone's confidence.

The Bakken has become a known resource play. Ninety-nine times out of a hundred, a Bakken well—wherever it is drilled—will predictably produce for decades. As long as oil fetches more than fifty or sixty dollars a barrel, the well will turn a profit. Companies can drill more than one well on a single surface location, or pad, producing certain economies of scale (the drilling rig doesn't need to be disassembled between wells, for instance). Drill. Change directions. Drill again. Last year, writer Richard Manning quipped in *Harper's*, "This is no longer wildcatting; this is plumbing."

The 1974 Safe Water Drinking Act regulates the injection of hazardous materials underground. It is the Environmental Protection Agency's best weapon for protecting our drinking water. In 2005, language was inserted that stated the term "underground injection" *excluded* "the underground injection of fluids or propping agents (other than diesel fuels) pursuant to hydraulic fracturing operations related to oil, gas, or geothermal production activities." This provision is known as the "Halliburton Loophole."

At the breakfast buffet in my hotel, two TVs are turned up loud, one on CNN, the other on Fox. Most men eat in silence. Biscuits and gravy, hash browns, eggs, bacon. Pink yogurt and a bowl of chopped fruit go untouched. There are lots of boots and overalls, many dirty from last night's shift. I heard people coming and going in the halls all night. What must be company men and management types tuck their shirts into khakis. It's Sunday morning, but everyone is rushing off to work. A man says, "I'm just saying it's like the Tower of Babel. You got everybody speaking all kinds of language. You just gotta speak English." His buddy responds, "Willie for President!" The crawl on the nearest TV mentions a toxin found in the drinking supply of Toledo, Ohio. Four hundred thousand people are without water. I catch a strong whiff of gas every time I turn on the tap in my bathroom, but it's probably my imagination. Then again, there is more than one well across the street. The talking head on Fox says, "People will make bad choices. You just have to let them learn the consequences."

Last December, Jacob Haughney, a laconic bearded guy with a pierced lip and a knit cap, posted a video to YouTube called "Check this out. Funky water in North Dakota." In the video, he deadpans: "Hey, what's up. My name is Jake. I work in the oil fields out in North Dakota and I just found out something about our tap water that we bathe in and use to brush our teeth and shit. And, uh, it's pretty weird. You lemme know what you think." He turns on a bathroom faucet and sparks a lighter next to the stream of water pouring down the drain. Suddenly he jerks his hand back as a column of fire erupts—*fwhoomp*—then flares out. The water continues running. He reaches back in with the lighter. *Whoosh*—it happens again. He laughs. "I know. Really weird, right?"

A drizzly Sunday morning, religious programs on most of the airwaves. I pull up to a neon-pink shack in the parking lot of a motel. Terse guys in trucks roll up to windows on both sides. "What can I get ya, hon?" asks the "babe-rista" at Boomtown Babes Espresso, who wears red panties/short-shorts and a bright pink bra under a see-through white tank top. (The stand's Facebook page posts pictures of women in fishnets, thongs, and Day-Glo outfits below suggestions like, "Come try a blended White Chocolate Butter rum mocha from Brandi . . . It's an orgasm in your mouth.")

Today's babe-rista is alone in the shack, and she's in the weeds. The line backs up in both directions. "I'm sure you get this all the time, but can I take your picture?" She poses for the camera, then goes grimly back to pouring. I hand over four dollars for the coffee, which is too hot and too sweet. Without my having to offer, she keeps the change.

The next day, I visit C Cups Espresso, which—though tamer in dress—nonetheless represents my last foray into erotic coffee. I pay three dollars for my large "D Cup" of drip. From my lowly Corolla, I can barely reach up to the window, which is clearly meant for the cabs of taller trucks.

After Alaska, North Dakota is the most male state in the union. In McKenzie County, which includes Watford City, south of Williston, the workforce is 77 percent male. On backpage.com the ads for Bakken escorts refresh themselves daily ("Upscale, classy, and nasty," "U pick UR price," "MADAM ROSE WILL BREAK YOU," "free dinner dates w appt 1 hr or longer"). Posts on Craigslist read "Miss homemade cooking/baking while working in the oilfields??," "Rent a Housewife," and "Bikini/Lingerie Cleaning." This summer, the Justice

Department's Office on Violence Against Women announced a $3 million special initiative for the Bakken aimed to increase the prosecution of violence against women, as well as to support the survivors of such crimes.

At one point, a guy walks through the lobby of my hotel with a shirt that reads, "Ask for the sausage," with an arrow pointing down. The woman at the front desk says, "That's a terrible shirt," as he walks right by.

The First Lutheran, Gloria Dei Lutheran, Good Shepherd Lutheran, and West Prairie Lutheran Churches of Williston are gathering at Harmon Park for the tenth annual Summer Sizzler for worship, food, games, music, and fellowship. A posted sign reminds everyone that there is no camping in the park. I sit at an empty picnic table in front of the band shell. Many people have brought their own chairs and umbrellas. Next to the PA system stands a large wooden cross crowned with sharp thorns. Behind us, they're blowing up the bounce castles, three of them, as the choir hallelujahs to the accompaniment of a piano, drum set, and red electric guitar. They sing, "We are all children of the highest," as the sun breaks through the clouds. The pastor takes the mic to ask everyone to please leave the bounce castles alone until after the service. He wears an orange T-shirt that reads, "God's work, our hands." We sing, "All are welcome in this place," as I am joined by two white-haired old women, whose pleasant perfume wafts my way.

According to the *New York Times*, one in six North Dakota wells had an "environmental incident" in 2013. A color-coded online map built by an independent cartographer shows the hazardous spills in North Dakota from 2000 to 2013. The northwestern quadrant is bathed red (oil), purple (saltwater/brine), and yellow (other hazardous materials). Many spills occurred along the banks of Lake Sakakawea, a giant Missouri River reservoir outside Williston that supplies drinking water to much of the state. A smaller cluster—of some four hundred spills—occurred in the southwest corner, just outside the town of Marmarth, where there was a short-lived boom before the Bakken took off.

I flip through the incident reports, full of stark language and times, dates, and coordinates. Most often, the contact information seems to lead to an address in Texas.

Today's gospel comes from Matthew, chapter 14, verses 13 to 21—Jesus and the loaves and fishes, a story of prodigious bounty, about having enough for all. The offering will go to world hunger. Little blond children come forth and sing, "Jesus made it more than enough to feed the hungry world." There is a skit with adults dressed as kids that involves a Darth Vader mask, a green light saber, mud pies, and a woman in pigtails; it seems to be largely ad-libbed and goes a little too long. There is joke about lutefisk, but the highlight of the skit comes when the "daughter" sings a silly song that goes, "Give me gas in my Ford, keep me trucking for the Lord." "Hallelujah!" some of us shout.

On September 29, 2013, a farmer outside of Tioga was harvesting wheat when he saw crude spewing some six inches out of the dirt. What he discovered was 20,600 barrels of oil spread across the equivalent of seven football fields—one of the largest onshore spills in U.S. history. The cause: a quarter-inch break in a twenty-year-old underground pipeline running to a rail yard. Cleanup could cost more than $20 million and stretch into 2017. So far, the San Antonio–based company responsible, Tesoro Logistics, has not been penalized. The farmer told the Associated Press, "We expect not to be able to farm that ground for several years."

As I listen to the familiar Bible story, I notice something new. The disciples start with five loaves and two fish, and—after feeding more than five thousand people—they end up with twelve baskets of leftovers, which is far more than they started with. Suddenly the verse seems less like a parable and more like a warning—there are times when you get much more than you asked for.

Since 1776—when recorded observation began—Youngstown, Ohio, had never had an earthquake. But one year after a well began disposing of wastewater from Pennsylvania fracking operations by pumping it deep underground, the town recorded 109 quakes. The seismic activity traveled east to west along a fault line leading away from the well, and coincided with disposal operations. (There were lulls during holidays, when the well was closed.) After a 4.0-magnitude quake shook Youngstown on New Year's Eve 2011, the disposal well was plugged.

Fracking and wastewater injection have been linked to thousands of quakes across Oklahoma and Texas. In the three decades before 2008, Oklahoma averaged two quakes a year of magnitude 3.0 or greater. By the beginning of 2015, it averages two a day. So far this summer, in the first half of 2014, Oklahoma is the state with the most earthquakes. (By year's end, it will see 585 quakes—about three times as many as California. The next year's count: 907.) In May, the U.S. Geological Survey warned a bigger quake—magnitude 5.5 or greater—could hit soon. Earthquake insurance is booming. Oklahoma currently has some nine thousand wastewater injection wells.

We offer prayers of intercession for all those in Iraq, Syria, Israel, and Palestine. We pray especially for "those in the oil industry who are far from their families." All are welcome to take communion; gluten-free bread is offered in the center of the three stations. I receive a blessing after dipping a hunk of bread into a chalice of wine held by a man with a giant belt buckle. The lines are long; things are a little informal outdoors. My neighbor, Marilyn, and I laugh as the band gets a bit jazzy with "Jesus Loves Me!" We strike up a conversation. She graduated from Williston High School in 1956; the pavilion is named for a beloved music teacher of hers who directed the city and high school bands for over fifty years. She wants me to know about the state's 125th anniversary celebration; she's pumped about that. She complains about the crowds, the construction, the dramatic spike in rents. But in the end, she insists, "I was born and raised here—I won't let them run me out!"

Last December, scientists at the annual meeting of the American Geophysical Union presented a study that suggested using fracking boreholes as a way to dispose of nuclear waste. Deep shale is impermeable to water. Theoretically, the heavy toxic slurry—which would remain radioactive for some hundred thousand years—would seep steadily downward toward the earth's core. Other scientists remained skeptical. Fracking fluids dumped into deep disposal wells have migrated back to the surface.

After the service, the congregation bounces in the castles (at last!), tosses beanbags, and eats hot dogs, burgers, chips, cake, popcorn, and cotton candy. A rockabilly duo sings praises to His name, but I leave before the main act, a western band. On my way to the parking lot, I watch some skate punks ogle the buffet. One gets halfway to the spread before losing his nerve.

The Fort Berthold Indian Reservation was created in 1870. Over the years, the Arikara, Hidatsa, and Mandan people—collectively known as the Three Affiliated Tribes—have conceded land, often under coercion, to the railroads, the government, and the massive flooding necessary to create Lake Sakakawea. One-twelfth its former size, the reservation now occupies a million acres in the western part of North Dakota and contains 414 wells on trust and fee land, which account for nearly a third of the oil produced in the state.

Over the recent Fourth of July weekend, a pipeline spilled twenty-four thousand barrels of so-called "saltwater" (a nebulous industry term for briny wastewater) onto the Fort Berthold reservation. Officials said the spill, which went unnoticed for some time, stretched nearly two miles, scorching vegetation and soaking into the ground. Many feared the spill had reached a bay that fed into Lake Sakakawea, the tribes' source of drinking water, but the EPA stated the spill had been stopped by beaver dams. A year later, radiation in the soil measures eight times the background level. Fracking saltwater can contain oil residue and chemical by-products and is estimated to be ten to thirty times saltier than seawater.

> And Abimelech fought against the city all that day; and he took the city, and slew the people that was therein, and beat down the city, and sowed it with salt.
>
> —JUDGES 9:45

In the Bible, Abimelech is an illegitimate tyrant of great ambition who usurps the throne by killing sixty-nine of his half brothers. As his power grows, he attacks his own city and slaughters his own people. In the end, he is dealt a mortal blow when a woman drops a millstone on his head.

"Bakken pollution catches everyone by surprise" is a headline that ran in the *Bismarck Tribune* on October 23, 2009.

In early 2011, in order to protect its state industry against federal environmental regulation, the North Dakota legislature introduced House Bill No. 1216, a nine-line, ninety-four-word emergency measure stating that "hydraulic fracturing is an acceptable recovery process in North Dakota." "Acceptable" meant legal—and safe. The bill passed the house and senate with a combined vote of 138 yeas to one nay.

"Where there is no vision, the people perish" is the motto that appears atop the banner of the *Williston Herald*. The front page also lists the

North Dakota rig count, which this morning stands at 194. Last week, the EPA opened a criminal investigation office in Bismarck, which it hopes soon to staff permanently. Before that, the nearest environmental detective was more than five hundred miles away.

I drive the streets of town. There are some lovely one- and two-story brick or wood houses with tidy lawns and maybe a new boat in the driveway. Other areas are not as nice, with chain-link fences, KEEP OUT signs, and busted windows.

On the east side of town, a man holds up a SLOW sign as trucks crawl past the entrance to what was a warehouse. A train whistle blows—we're next to the tracks. Men in hard hats and orange vests go about their business among blocks of rubble, burnt barrels, bent steel, the listing skeleton of a building, twisted heaps of who-knows-what. A charred tower looms overhead. An orange windsock blows perfectly horizontal; I breathe in a wet-charcoal smell laced with a strong chemical tang. Something smolders inside the pile, drifting smoke across the ruins. I peer over a flimsy fence at pools of oil-black water. From a fire hydrant across the street, a thick yellow hose snakes beneath the fence, leaking water onto the pavement.

Two weeks ago, just after midnight, the Red River Supply storage facility exploded into flames. The smoke rose hundreds of feet into the air, and though the wind was blowing away from town, a half-mile area was evacuated and the FAA delayed flights for six hours at the airport. The company stored dozens of chemicals, including drilling fluids, soil and dust products, and proppants, most of which burned. It took officials more than six hours to alert the public of any possible danger.

Two little kids with buzz cuts ride their bikes past me. It's hot; the blue bulbous Williston water tower seems to shimmer in the distance. The chubbier kid lags behind. "Where are we going?" he whines. His friend replies, "Up your ass and around the corner!" as he speeds off into the rec center parking lot.

Amenities offered at the Williston Area Recreation Center, a $70 million, 234,000-square-foot angular and soaring steel-, stone-, and glass-sheathed complex that opened in March and is said to be the largest such rec center in the country: four tennis and basketball courts, a turf soccer field, batting cages, a two-hundred-meter track, an Olympic-size swimming pool, three birthday rooms, an indoor playground, child-sitting six days a week, meeting rooms, a patio, an overhead track, a weight room, a cardio area, a spinning room, two water slides that loop outside of the building, a "teen area" (TV, pinball, table hockey), a golf simulator, a kiddie pool, an instructional kitchen, an ice rink, racquetball courts, and a water park, which includes a lazy river, a surfing simulator, and an oil derrick spouting water, beside which dance three playful orange-and-yellow flares.

Fracking the Bakken releases both oil and natural gas. Natural gas is cheap and not easily shipped—given the lack of pipelines—and so companies focus on the oil, which can be moved by rail, while burning off the gas at the wellhead. In other words, it is cheaper to waste the gas than to sell it. The process is called "flaring." North Dakota wells burn off about 30 percent of the gas they capture—far, far more than wells in the rest of the country, which flare less than 1 percent. (The worldwide average is believed to be less than 3 percent.) In April, North Dakota wells burned off $50 million worth of natural gas—enough to heat more than 1.5 million homes—and more than twice the amount consumed in the state that very month. Such flaring releases more CO_2 than a million cars would in the same period. Companies can flare for a year without paying taxes, after which a waiver can be applied for (and usually is granted). The state loses about $1 million a month in gas tax revenue. At the beginning of July, the state announced new flaring targets—only 23 percent by next January, dropping to 10 percent by 2020.

In a YouTube video called "Lighting a cig in Williston," uploaded in 2012 and seen a mere 214 times, one guy holds the camera and laughs moronically as another guy in sandals and shorts shambles over the wall of an earthen pit and approaches a roaring flare.

> CAMERA GUY: We're lighting a fucking cigarette off a fucking flare.
>
> SANDALS GUY *(whining)*: There's no way—it's so hot.
>
> CG: DO IT! Pussy! *[He giggles as his "friend" leans closer to the flare.]*
>
> SG *(shuffling back a few steps)*: Ah, it's so hot!
>
> CG: Did you get it?
>
> *Suddenly, the flare explodes, sending a fireball rolling across the pit. SG makes guttural noises and retreats toward the camera.*
>
> CG: Oh, fuck! *[Laughs maniacally. The camera shakes to reveal he's sitting in the driver's seat of a truck.]* GET IT!!!
>
> SG: No! *[Unintelligible grunting.]*
>
> CG *(wheezing)*: Oh, fuck.
>
> SG *(scrambling out of the pit)*: I swear they just turned that on full blast.

My first night in town I saw them from the highway: twin pillars in the dark, flickering brighter than any other lights on the prairie. I followed a dirt road for about a mile before coming to the two flares. One burned taller and with a bright-blue base. Some flares roar like jet planes, but that night the sound—a loud *whoosh*—came in waves: stronger, as the fire burned brighter and higher, then receding and softening. A chorus of crickets tried to sing above it. Mosquitos swarmed, but I stood and listened. I thought of the view from space. I'd come to see this. The noise reminded me of something I couldn't put my finger on. Later, trying to fall asleep to the Doppler whine of trucks, it hit me—the flares sound just like the sea.

Seagulls soar above the Walmart parking lot—curiously out of place, as if they've lost their way like the rest of us—as I inch my car through the throngs of shoppers pushing carts filled with water, ramen noodles, sports drinks, and the like, a prodigious bounty of quick-fix nutrition. A car alarm goes off as I pull past a sign that reads TOWING ENFORCED AT ALL TIMES. Two years ago, Walmart kicked out everyone living in its parking lot, dozens of men in tents, cars, and RVs, some of whom had been receiving mail at 45 Walmart Parking Lot, Williston, ND.

To some, Williston represents the freest of markets, unbridled supply and demand, an anarcho-capitalist dream. If prices are high, so are the wages. In 2012, entry-level jobs on a rig paid about $66,000, while oil-field workers earned, on average, $112,462 a year. This summer, staff jobs at Walmart start at $17.50 an hour, nearly two and a half times the federal minimum wage. On the day I visit, I pass a man holding a cardboard sign reading HELP ME—U.S. VET.

Just some of the jobs advertised in the *Shopper,* Williston's ad pages, on July 31, 2014, an issue that stretches seventy-six pages: hot shot driver, corrections officer, building inspector, heavy equipment operator, staff engineer, social worker, water resources technician, diesel technician, lot person, dispatcher, belly dump driver, workover rig operator, pusher, carpet cleaner, janitor, pumper, roustabout, saltwater disposal operator, shop hand, chemical yard driver, receiving clerk, bookkeeper, insurance agent, collections specialist, drilling superintendent, machinist, substitute teacher, seventh-grade football coach, activity aid, slick driver, winch driver, scoreboard operator, service hand, and game official (all sports).

I get to the library right at 2:00 p.m., when it opens. A card thumbtacked to the lobby bulletin board lists the red flags of human trafficking. ("Has an adult boyfriend, owns expensive items they could not afford, has bruising or branding tattoos, uses slang like *wifey, daddy,* or *the game*.") There is a number to call. To the left of the circulation desk, a dozen men stare blankly from the wall—color mug shots of local "high-risk" sex offenders (with addresses and aliases). The library is quiet, just the drone of the lights and the tapping of men browsing the internet. Guys on laptops and phones are spread around the periphery, using the free Wi-Fi. A starburst quilt hangs on the wall. I wander past the new books to a shelf of local newspapers. I scan the headlines: "'Isolated' Incident: Bacteria Found in Water Samples"; "1 in 7 Private Sector Jobs Tied to Oil, Gas"; "Missing Girl Found Safe in Williams Co."; "NDSU to Fund Work to Reduce Dust"; "Housing Booms as Needs Grow in North Dakota Town."

I trade my driver's license to a young librarian for a copy of the *Williston Herald*, which is kept up front. I have to wait for today's paper. When it's finally my turn, I read the lead story on the front page, which begins, "A thirty-eight-year-old Williston man is being held on $2 million bond after being arrested for two separate cases involving human trafficking and sexual imposition."

"Charged with a serious felony? You need serious help!!! The job of the State's Attorney is to get a conviction. Shouldn't you have an aggressive and experienced legal professional protecting your rights? Our office has helped countless numbers of people stay out of jail and move on with their lives . . ." begins one print advertisement I saw in Williston.

While I read the paper, a man approaches to sign up for a computer. The young librarian tells him the wait will be an hour and a half. "Already?" he asks as he puts his name on the list. The library has been open for thirty minutes. The young librarian leaves him to make her rounds warning the men who have five minutes of computer time left. She wears moccasins and is polite but firm. You can only use the computers once a day. Guys keep adding their names. A woman in rainbow tie-dye rails against the library's policy of not giving cards to people living in trailers. She has her daughter with her. The librarian remains calm in the face of mounting frustration. The woman slaps something down on the counter. The librarian sighs, "I'm sorry, honey. But this ID is for West Fargo." On the desk sits a thick binder of local job and housing information, from which a guy pays twenty cents to photocopy a page.

Later, a librarian explains to a man how to apply for a card. When she's done, she helps another man at a computer terminal find some tax forms online. Next he tells her he wants a map of Texas. He speaks quietly in broken English.

The librarian says, "You want to print a map?"

The man nods.

"Are you looking to get there?" She raises an eyebrow. "From here?"

Again he nods, and they switch places. He fidgets in a long red-and-white striped shirt that nearly reaches to his knees, as she prints driving directions from here to somewhere, anywhere in Texas.

Leaving the library, I see the man standing in the middle of the parking lot looking lost, no car in sight.

A sunny afternoon in the park—the skate punks are out in droves. Tough kids, scrawny kids, they hang in, around, and off of a beat-up blue truck. One in a bandanna swings a yo-yo. Most wear ripped black tees, though some go shirtless. They saunter up and nod, many without boards, just a tribe checking in. Mothers don't give them a second look as they lead screaming toddlers past them to the parking lot. There might be some shady stuff going on behind the dumpster.

Two disheveled fortysomething men sit in the central pagoda, surveying the park. One guy does push-ups while his buddy heads to the bathroom. Like me, they have nowhere to be. I feel awkward just sitting in the park, given the mug shots in the library.

A black truck drives up, and the skate punks flock over. I approach two that hang back, a guy and girl sitting atop a ramp and petting a small red dog between them. She has blond hair streaked pink and a nose ring. He has short sandy hair, dark jeans, and a black Red Hot Chili Peppers T-shirt that I almost find hard to believe. They've lived in town for three years. I ask how it's going.

"A lot of kids don't like the new kids," the girl says. "They're mean to them."

Trying to be helpful, the guy adds, "You know, the changes and all." Public-school enrollment increased 12 percent in the past year.

The girl says, "But now that I've been here three years, it's okay. There are some nice people in this town, I guess." Behind her, someone bangs on the hood of a truck.

I ask, "So do you like living here?"

The kids let the question hang for a minute while the guy picks at his arm, before giving me the answer of any teen, anywhere. "Fuck no."

The week I left for North Dakota, a Williston man pled guilty to tying nooses around his ex-wife and son and dragging them behind his SUV. Reportedly, he meant to snap their necks, saying, "You know what I do with liars." The boy asked, "Mommy, are you going to miss me?" After a short distance, the man's current wife put a stop to the shenanigans. The couple recently had been arrested on a meth charge. Two days later, there were reports that two men in a black Silverado fired a shot at a seven-year-old girl on her bike, who escaped with minor injuries and a flat tire.

I head to dinner in my hotel cafeteria. As I go down the line, I ask for shrimp, fettuccini Alfredo, green beans, and corn bread. The server asks, "That all?"

One breakfast I saw a guy load his tray with an everything omelet, biscuits and gravy, sausage, bacon, two pieces of pecan pie, and three slabs of French toast. He told those of us in line, "I was working all night keeping the lights going. It's time to get my eat on!"

DUE TO OVERWHELMING RESPONSE WE'RE BACK TO DETECT YOUR RISK FOR STROKE IN LESS THAN TEN MINUTES reads a flyer from an outfit in Iowa that was going around Williston when I was there.

The room is loud with guys scarfing down food and chatting about work. The table behind me discusses pipe fittings, before talking shit about someone named Sweeney who moved back to "Colorado or Alabama or Texas or something."

One man says, "He didn't know shit when he put in that pump jack."

Another adds, "We could have saved everyone some time, put that shit up right."

The third gets the last word. "What we have there now is an old-fashioned clusterfuck."

The men sit in groups segregated mainly by age. No one sits with me. Someone shouts, "Don't tell me karma's not real. That shit's real as fuck." He then discusses a playful fight between two guys named Big Steve and Little Steve, which started with someone shooting a rubber band and ended up with a screwdriver in the eye.

Monday evening in the hotel lounge, which consists of four round tables and as many stools before a bar. A guy walks up and tells the bartender, "I took my laundry somewhere else, since you wouldn't do it." She laughs and says, "I've got eight loads already." She wears glasses and a ponytail, and has a tattoo on her forefinger just past her knuckle. He says, "Okay, I'm gonna take a shower and come down for a beer."

These men are doing difficult and dangerous jobs. Oil and gas workers are 7.6 times more likely to die than workers in other industries—and oil and gas workers in North Dakota are more than six times more likely to die than roughnecks across the country. The state has the highest rate of on-the-job deaths in the U.S., more than five times the national rate.

BARTENDER: What'd you do today?
DEEP SOUTHERN MUMBLE: Fixed a tractor and got all these boys stocked up for two weeks.
BARTENDER *(singsongy)*: Because you're getting ready to go!
MUMBLE: I'm gone in the morning.
BARTENDER: Hallelujah.
MUMBLE: I just hope they don't fuck it up while I'm gone.

Catastrophic well accidents—blowouts, getting crushed or caught in machinery, etc.—are only part of the worry; occupational hazards abound. A toxic oil-based swill known as drilling "mud" cools and lubricates the bit and helps remove the cut rock; exposure causes lesions. Fracking fluid usually contains a number of carcinogens. Workers involved in the flowback operations have been shown to have high levels of benzene in their urine. Wastewater pumped from the well can emit hazardous concentrations of hydrogen sulfide, a lethal, often-odorless gas. Noxious vapors waft from tanks and open-air pits. In 2012, a North Dakota oil worker was found dead on a catwalk above a storage tank, poisoned by the concentration of petroleum vapors. (In a statement, Marathon Oil said, "Our analysis and testing of the location following the incident indicated no apparent equipment malfunctions or other abnormalities." The company settled out of court with the man's family.)

"They got the string fucked up. Sent the wrong part from Minnesota. We're only like 150 feet from bottom. The whole thing's already 320 joints."

"You get the rig fixed?"

"Took all fucking day."

"Still getting paid."

"Yep."

"I'll drink to that."

Two years ago, after conducting a study across five states, federal officials cited another danger: inhaling the fine sand used for fracking. "Frac sand" can cause the incurable lung disease silicosis, as well as cancer. Air samples at wells consistently exceeded healthy guidelines; nearly a third of the samples were ten times over the occupational limit.

At a table behind me, four older, dressier guys have been drinking round after round of doubles—at least six, by my count. Crown Royal, Captain Morgan, well vodka, and whiskey mixed in an assortment of water, orange juice, Coke, and Diet Coke. At least one guy has a German accent—a bunch of international good ol' boys. One of them shouts across the lounge, "Hey, Kruger, don't let her put too much ice in them drinks."

A man with a pierced lip has had it with Cedric. "He's gonna find out just what kind of asshole I am," he tells me. The bartender sighs sympathetically: "Cedric, Cedric." The guy can barely stand it; he's practically vibrating. "I'm gonna cut his hours down to eight a day, five days a week. See how he likes that!" How many are they working now? "Twelve to fourteen a day and it's going to go to seven days a week." He shakes his head. "Me in my forties, the other guys in their fifties. Shit. You got old-ass fuckers who can outwork all them youngsters." I nod, and he says, "They come in at four in the morning. Four a.m.!" I don't know exactly what he means—should they be going to bed sooner or getting to work earlier?—but we're full-on ranting now: "The other day Cedric was drunk as a skunk. I'm gonna start running that shop like Joseph fucking Stalin!" The old roughneck sits down, deflated. "Okay, shit, just a fast beer," he tells the bartender. "I gotta work tonight. Three-thirty is gonna come pretty damn quick."

For about a year, the bartender has been spending six weeks at the hotel, then two weeks back in California, where her mom lives. She confides that in three months, "give or take," she'll be moving to either Dallas or Fort Worth, where she has more family. The hotel has become part of a national franchise, which is bringing in its own management team. "The owners don't want anything to do with the place—they just want to stay away and collect their money." A lot of people are quitting or leaving. "Change is coming," she assures me.

In the stairway, an older man is hunched over his phone, mumbling, "Hey, hey there. I hear you, brother. It's *hard.*" I step around him. His eyes are closed; he doesn't look up. I've noticed a lot of guys on their cells in the parking lot; either they need to smoke or get air or they're sharing a double and want a little privacy. Meanwhile, the room across the hall seems to be having a party. I hear guy's voices, hip-hop, and the sounds of PlayStation.

One day I walk the downtown in all its boomtown contradiction. Around the corner from the sushi place (where "a touch of California atmosphere almost makes you forget you're in Williston") stands an old quilting supply store. Next to a hair and tanning salon, a sidewalk sign offers GUNS & AMMO. Under a peeling awning, bridal dresses line the dusty window of a store that also advertises 50% OFF LAWN AND GARDEN. An upscale cooking store, some junky antiques, an ad for hot stone massage, windowless bars that look like bunkers, a Hallmark store.

The retail scene is rather sedate at the one-room JCPenney department store, which opened in 1916 and today is offering early back-to-school specials, despite its older clientele, a couple of whom power-chair themselves along the aisles. I buy a shirt that's been marked down five bucks. The clerk looks a little bored and uncomfortable in a turquoise button-down and a white-and-black striped tie. He's in high school and says, "You got an accent that's new to me—where are you from?" He moved here from Mississippi with his stepdad, and his accent is no slouch, either. He's been here a month and likes it so far. "I was real glad to get this job," he says. "Where I come from, they pay $7.25 minimum wage, so $10.26 sounds pretty great to me."

Later I see that one of those windowless bars—which, from the outside, seemed too seedy to enter—is for sale for a cool $1.5 million.

They're tearing up the bottom of Main Street, so it's tough getting to the train station on the south end of town. An earthmover sits motionless in the middle of the road. The streets are once again dirt, like a real Wild West town. A sign hangs over the boarded-up K.K. Korner Lounge, which online comments swear is named after the owners and not a hate group. Everyone knows the town's two strip clubs down by the Amtrak station—Whispers and Heartbreakers—neither of which I can bring myself to enter, despite the fact that the latter is getting a new coat of white paint. These are the first establishments you see when you step off the train. Next door is the No Place Bar, which welcomes bikers and today has a pink-and-black baby stroller abandoned out front. On the side of the Salvation Army, across a small parking lot, someone has put up a giant billboard of the Ten Commandments.

Between 2008 and 2012, the number of cases of gonorrhea in the western half of the state rose 72 percent. Chlamydia was up 240 percent.

One o'clock in the afternoon and there's a guy sleeping in the shade of a stunted tree outside the train station. The landscaping smells of urine. In the small waiting room, a TV plays loudly as a train scrapes down the tracks. They sit on seats and on their bags—young, old, with kids, without. They're waiting to head west on the Empire Builder, which is already an hour and a half late. This summer, oil trains have been wreaking havoc with Amtrak's passenger service, with delays averaging three to five hours and sometimes stretching into the double digits.

Around Williston, the homeless population hovers around one thousand; there is no city shelter. The town is prioritizing infrastructure, such as roads and sewers. The Salvation Army has bought hundreds of men bus and train tickets back to where they came from. The town has a newly elected mayor. Last winter, as a city commissioner, he opposed a church organization's plan to use a National Guard armory as an emergency shelter. When he announced he was running for mayor, he told the *Williston Herald:* "My approach is: the number one priority is the people that live here."

Higher up Main Street, on a door leading into a newer-looking but nondescript building, a handwritten sign announces that the Bakken Club will be closed for breakfast and lunch this week and won't open for full service until four o'clock. That's a few hours away, but it's not like I could eat there anyway, or take advantage of its wine list, airport shuttle, event space, business center, or plan room (where I could store my maps and blueprints), because, according to its website, the Bakken Club is "a limited-membership Social and Business Club, Bar, and Restaurant for individuals with a long-term, vested interest in the region"—which certainly isn't me. I'm a transient, just passing through, so I'm left peering in at the swinging saloon doors, deer-antler lamp, and Andy Warhol–esque silkscreen of a cowboy Clint Eastwood.

Open-mic night at J Dub's, one of the local spots, where trophies from state dart competitions hang above the bar and I can see five flat screens from my seat. A sign on my table advertises stand-up comedy every Thursday—I am five weeks too late to see Dustin Diamond, a.k.a. Screech from *Saved by the Bell.* A twentysomething shuffles up. The bartender says, "Ready for the day to be over?" The guy nods and she talks him into some elaborate shot of at least four ingredients. He downs it and wanders into the back, where they're setting up the open mic, about which my table ad suggests: "Leave your heart on the stage." I ask the waitress if the entertainment is any good. She gives a conspiratorial shrug: "Depends what you like," implying the negative. Then she laughs. "Naw, you never know."

In the back, a bearded guy in a newsboy cap is busy getting ready. He drops a cymbal, and the audience laughs, surly. He must be the emcee. White shirts glow under the black light. A waitress appears in a short polka-dot dress, tattoos crawling up her shoulders and neck. George Strait is playing, loud. Forty-five minutes pass and the guy has assembled a drum kit, that's it. A country cover of Queen's "Fat Bottomed Girls" comes on. The man in sunglasses at the table next to me is unexpectedly moved. He bellows, *"Chuuuug it!"* His friends don't respond. The emcee sits at the foot of the stage, scribbling in a notebook. The flat screens show preseason football and Ultimate Fighting. Everywhere, men beating each other up for money.

Last December, in a late-night scuffle, a thirty-one-year-old man from Mississippi died after being run over in the parking lot of J Dub's. After spending six days tracking down the driver, the police searched his cell phone and found a text from one of his passengers: "Dude, that guy passed away last night. WTF we should really do something. I feel terrible."

At breakfast, one of the cafeteria TVs is tuned to cartoons, though there are no kids in sight. Multicolored dinosaurs are dancing as they ride a train into the past.

I go to check out, and a guy in a red sweatshirt is talking to the receptionist. It's 9:30 in the morning, and he has a beer in his hand. When I approach the desk, he tells her, "I'm going to go see where the boys are."

I'm surprised at my bill. As I reach for my wallet, I realize that I, too, am part of the boom.

My drive to Watford City takes an hour and a half, a forty-five-mile death march of traffic through nearly constant construction. I pass a

Dodge Ram with what looks like a giant tractor tire tied to its roof. Two days earlier, I was driving outside of Williston when suddenly a truck tire bounded across my lane. I swerved, going seventy, and the tire hurtled into a ditch. Ahead, a blue Silverado was pulled into the median, bare axle showing. Leaving town, I saw the truck still slumped by the side of the road. Yesterday, I watched a cement mixer abruptly start lurching across two lanes of highway, forcing everyone to slam on their brakes. No one even bothered to honk. Signs on the highway exhort PASS ON THE PASS.

The flares are big this morning; more rigs are going up. RV cities sprout in fields, the vehicles packed tight, nose to nose. On the outskirts of Watford City, a man camp marches its rows of trailers into the prairie. I keep my eye on a mean storm brewing in the distance. Weather here is no joke—the hail can kill a calf. On Memorial Day, a tornado ripped through a nearby trailer park, injuring nine people. Two oil workers shot a harrowing video that made the national news. The 120-mile-per-hour funnel passes within one hundred feet of their car, where the men eventually seek shelter. They laugh helplessly while the twister blots out the sun. Debris smacks their window as the funnel sits above them, spinning an angry red. One of them giggles, "Dude, where do we go? We've got nowhere to go!"

How to find flares in Watford City: Take Main Street past the big apartment blocks going up, still wrapped in Tyvek, and follow the road until it dead-ends at a dirt intersection. Wind the empty grid of rutted clinker roads that leave your tires red, stair-stepping closer and closer to the flames, many of which are visible for miles. Watch out for the enormous trucks lumbering past—they'll edge you over to the side, where sharp rocks might pop your puny tires. Pass new-looking yield signs posted at the dusty intersections of barren roads with long, hopeful names like 128th Avenue NW. When you're done looking at flares, pull up to a rig.

The storm is off to the north; the wind is blowing in. I smell wood smoke and burning plastic. Towering over the prairie, the rig is orange, blue, and white and lit brightly even in the day. A field of hay bales rings the well pad. The smell of gas is strong, like when you spill some at the pump, but far more powerful. There are rows of tanks and lengths of pipe. Something emits a high humming sound. Men come and go on the platform. No one looks my way. A large sign gives the name of the company and a three-digit rig number. After watching for a while, my head begins to hurt and my throat feels gummy. I get into the car and roll up the windows.

Watford City celebrated its centennial about a month ago, but it's only recently come into its own. The houses are more modest, a little shabbier than in Williston. This is a much smaller city, more boots and cowboy hats on Main Street, more unpaved roads in town. But developments are cropping up on the fringes: off the highway, to the south, and over by the airport (which is slated for a $2 million expansion). Construction is everywhere. A $59 million medical complex has broken ground. The First Lutheran Church is nearly done adding a big wing. A new high school will be built this year, next to a planned $57 million events center. At the current high school, home of the Watford City Wolves, a colorful electric sign announces new-student registration for grades six to twelve started yesterday at 9:00 a.m. A girl crosses the parking lot with her mom, who is carrying pieces of paper. In Watford City, one in four children is homeless, meaning he or she fits the federal definition, which includes those living in a hotel or RV. That comes out to 263 kids.

The city's official population is close to twenty-eight hundred, but this summer the police department estimated that temporary residents pushed the actual number to somewhere between six and seven thousand. In June, the mayor of Watford City told the *Bismarck Tribune*, "It's not good for families to live in RVs."

Five miles south of town, just off the highway where you see the grazing horses, down the dirt road, and past the toddler scrambling up the hill and the trailer offering ten-dollar hot and ready pizza, they rise before you like some postapocalyptic city, a gathering of the faithful ready to make the final stand: eight insulated, climate-controlled hangars, each holding eight bays with room for three RVs per bay—a total of 192 vehicles—plus outdoor pads for another seventy RVs, some of which might be on the waiting list to move inside. The brainchild of two Minnesota builders, the North Dakota Indoor RV Park offers tenants electric, sewer, water, and propane hookups; halogen lights; fire and carbon-monoxide detection; a hanging electric heater; laundry service; and two parking spaces outside each overhead door. Rent runs from $1,000 to $1,200 in the warmer months, depending on the size of the bay, plus an extra $250 a month in the winter. Outdoor spots cost $750 year-round.

The morning I visit, most bays are closed—everyone is at work. The speed limit is five mph. A sign says SLOW, CHILDREN PLAYING, but I don't see any—just a bald guy walking a poodle in the gray morning drizzle. Satellite dishes sprout from the sides of the hangars. The road ends in front of big yellow dumpsters. A food truck wrapped in an ad for a drilling-pipe company is serving some guys coffee. The office is open, but no one is behind the counter, on which a notice reminds people to sign up for the group email. The common building offers a couch, some seats, two arcade games, two pool tables, a treadmill, a bike, a dartboard, a kitchenette, and some vending machines. A TV plays *Scooby-Doo,* but the hangar is empty. Lined face-up on a wooden table between the mailboxes are a Jackie Collins romance novel, a thriller by David Baldacci, and two copies of the King James Bible.

Later, when I call, the guy tells me that they're full right now, but he can put me on the waiting list. "Things open up all the time," he says, hopefully. "People come, people go."

In January, a Louisiana man took shelter in a Watford City dumpster. A few months earlier, he had biked down from Williston with his last twenty dollars. That night, he burned two blankets, his boot liners, and a scarf, but it wasn't enough. He burrowed under trash as his body went numb. Two and a half days later, he managed to crawl out and get himself to the gas station nearby. Someone drove him to the emergency room. He awoke in a Minot hospital with both of his legs amputated below the knees.

Overall, the state estimates that some 45,463 people are living in temporary housing in the western half of North Dakota. That comes to one in five people.

This month's deposit into the state's Legacy Fund—which collects 30 percent of the oil and gas tax revenue and can't be accessed until June 2017—was the largest so far: $112 million.

In April 1886 at a creek southeast of town, Theodore Roosevelt caught up to three thieves who had stolen his boat. The president-to-be put them under citizen's arrest and walked them for thirty-six hours through the mountains to face the sheriff in Dickinson. All over town I keep seeing Watford City's own sheriff and his muddy white truck. His fellow officers seem to be out in force, too. I watch them pull someone over. The man does not look pleased. In 2009, Watford City had eighteen arrests for DUIs. Last year, they had 254.

That night, I will spend an hour trying to find dinner. The girl at the pizza shack won't sell me anything ("Oh you don't want to wait. It'll be a long time"); a table for one at Outlaw's Bar & Grill will be at least forty-five minutes. I'd sit at the bar, but it's packed. Across from my hotel, Outsider's Bar & Grill will be similarly swamped. In the first quarter of this year, taxable sales in Watford City were up more than 31 percent.

The Watford Hotel is swanky, for the region. A lot of business seems to be getting done here, as evidenced by the number of Bluetooth headsets and the guy in the men's-room stall who was on a conference call. The bar boasts plenty of dark wood, plus paintings of buffalo and cowboys. Patrons wear jeans and blazers, pearls and polos, a pair of shiny wingtips, golf-wear from a Flagstaff club. A well-heeled woman of a certain age drops a hundred on a thirteen-dollar bill.

A bunch of junior-executive types crowd the bar: Andy, John, Matt, and JT. They're getting drunk in a jolly, enthusiastic way. Andy—belted khakis, preppy plaid shirt, glasses, hair thinning, in his thirties—is their leader, at least in volume. I peg them for analysts in town checking on their investments.

"Were you on that call?"

"Oh, yes, we had some issues with that drilling manager, but we addressed those."

JT orders a margarita on the rocks. Everyone discusses his frequent-flier miles. Andy is the big winner, natch. The new guy tries to suck up: "Andy, got any interesting projects?" which leads Andy to opine on the fate of the company, vis-à-vis his place in it:

> The biggest thing at the company is revenue. So naturally I get a lot of attention. . . . We're constantly asking ourselves, "Where do we want to be, and how do we want to grow?" . . . It's all revenue and sales, revenue and sales . . . synergy . . . focus on your niche—*that's* how you gain market share . . . Brant and I were in Louisiana last week having dinner with a client, and we straight up told him, "We have the most sophisticated systems in the business. You know that, we know that" . . . Low-hanging fruit, that's what I tell them. . . . Who should run sales? Certainly not Larry! . . . *Teams,* that's the key. . . . What Brian said was so profound: "The one who does best, that guy should be in charge." And he wasn't even talking

> about me! But he was right. . . . And *that* tells the whole story with regard to our revenue issues. . . . Who owns the customers? Nobody owns the customers! But you've *got* to own them . . . it is what it is what it is. . . . In a week, I knew more about their business than they did. . . . The only thing they do is suck up some oil and sell some drums. Whoop-de-do. Bunch of small-time operators. . . . You've. Got. To. Figure. Out. What. Part. Of. The. Branding. You. Want.

Finished with his monologue and his beer, Andy asks the bartender if she has Rose's Lime. She does. He orders a gimlet—"Do you know what that is?" She doesn't. She works at two bars. Her brother is a mixed-martial-arts fighter, so she's keeping one eye on the TV, which, like most oil-patch bars, shows a UFC bout. "It's okay," Andy loudly assures her. "Just give me Ketel One on the rocks with Rose's Lime." Andy's protégé learns fast. He leans in: "Can you make me a mojito?" The bartender's face is blank. Then she tells him, "No mint."

Andy keeps at it. He asks for a filbert in his drink, which he says is how Errol Flynn drank them "back in the forties." The bartender just ignores him, but someone else takes the bait: "What's a filbert?"

Andy and his crew work for an environmental company. They're HAZMAT guys. They have a landfill four hours from here, but they do business in Wyoming, Texas, Colorado, all over. They service the rigs, dispose of drill cuttings, handle salt control, and so on.

As the night wears on, the guys argue over who is picking up the tab. Andy likes to pay for a stranger's drink, saying, "One for me and whatever my friend is having over there." The friend often doesn't know he's a friend. That's how Andy starts talking to a guy in a T-shirt, a Minnesota hauler who owns his own truck, a belly dump operating out of Dickinson. The guy says driving is a lot of hard work—too much, maybe. He's older than Andy, who starts to explain the ins and outs of his own business, saying, "We have a lot of fixed assets." The trucker says, "Uh-huh."

Andy offers to set him up with a local guy named Brant, because Andy, it turns out, lives in Massachusetts. He gives the trucker his card. As the trucker decides what to do with it, Andy turns his head and announces to the room, "Maybe I'll have one more and then cash out."

And that's how the boy kings of the Bakken talk when they're sitting at the bar.

A few miles outside of Killdeer, a bluebird flies into the hood of my car—*thwack!*—and flips up over the roof. In my rearview, I see it lying motionless on the highway. I ponder this omen as I roll up and down hills, passing empty grasslands, flares, falling barns, a pad with three tall wellheads and another rig going up nearby.

N.D. OIL ATLAS SOLD HERE is a sign on the door of one of the gas stations in Killdeer, which is also running a clearance sale on their "I love crude women" T-shirts ($9.99). Meanwhile, the billboard in front of the high school (home of the Cowboys) reads NOW HIRING. BUILDING MAINTENANCE. BUS MAINTENANCE. $60K PACKAGE.

On the way down to Dickinson, the rain picks up. Off to my right, a sign: PUBLIC SALTWATER DISPOSAL. Later, closer to town, I pass another disposal site—rows and rows of tanks. Somewhere, thunder booms. In the past two months, three saltwater storage tanks have been hit by lighting, leading to fires that lasted for days. The corrosive water is stored in tall fiberglass tanks—which last longer than metal, but also burn better. The saltwater itself is a combustible stew of brine, oil residue, and gas vapor. The most recent fire, next to a truck stop between Williston and Watford City, spilled 118,146 gallons of brine and spilled or burned 27,258 gallons of oil. There are more than 440 such saltwater storage locations across the state.

Dickinson, officially a city of some twenty-one thousand, is roughly the same size as Williston, though of course both figures would balloon if you added temporary residents. The number of babies born in Dickinson has tripled since the boom began. Jobs on the night shift of McDonald's come with a $300 signing bonus. My WPA guide to North Dakota from 1938 assures me, "The town is still young enough to retain much of the friendly atmosphere of the early West."

A Craigslist ad from two weeks earlier, titled FREE ROOM (DICKINSON):

> *Free room for casual hook up. All utilities and most food will be covered. Good sized bedroom. W/d in unit, attached garage. I'm a average looking 24 year old who owns his own home. Must be a single woman between the age of 18–30. Send photo with reply. Please.*

On the west side of town, I walk past condos and attached one-story new-construction homes with mud-caked trucks in the driveways and garbage bins pulled to the sidewalk. A two-bedroom, two-bath unit in one of the condos on this street is listed for sale at $205,000, below average in this town. The condo offers thirteen hundred finished square feet. Two girls ride a big wheel up and down the sidewalk. This seems like it would be a great neighborhood, a place of upward-trending dreams, except that directly across the street, behind a chain-link fence and a few struggling and widely spaced saplings, none of which have leaves, a pump jack vigorously nods its head up and down.

After three years of monitoring air pollution around drilling sites, the Colorado School of Public Health found that people living within half a mile of a fracked well were at an increased risk for cancer. Last fall, the chairman and CEO of ExxonMobil, Rex Tillerson, joined a lawsuit opposing the completion of a water tower that would enable fracking in his neighborhood outside of Dallas.

Every time I drive under the tracks that stretch over the main street in Dickinson, there is an oil train parked overhead. The black cars just sit. After a few days, it begins to feel like there is a single massive train stalled to the horizon. (I look, but cannot see an end.) All over town—in my car, in my hotel, wandering the streets—I hear the train blow its long, loud whistle. I learn to read the cars. A white stencil might identify the DOT-111 tank cars as belonging to a leasing company from St. Charles, Missouri, or Littleton, Colorado, or Chicago, Illinois, or Hamburg, Germany, but they could be rented to anyone and they could be going anywhere. They all bear the red diamond-shaped 1267 placard—petroleum crude oil on board. I become something of a rail fan, but the trains don't move, and so I'm left watching a trash fire burn beside the tracks. When I sleep, I see the long black worm wake and start to crawl down the line.

Hidden behind a green grassy hill southwest of town, the Bakken Oil Express terminal connects to the southern main line of the Burlington Northern Santa Fe. Oil trains run in twin loops, curling under a road and around earthen berms through some 55,000 feet of rail. In one corner, large white drums painted with the green BOE logo store up to 640,000 barrels of crude. Dusty tanker trucks lumber down the dirt road past the sign to the Dickinson Trap Club and into one of twenty-two bays, where their contents are siphoned off. The facility can load two trains at once. A sixteen-inch pipeline stretching thirty-nine miles north to Killdeer just came online last month. It can deliver 165,000 barrels of oil—the equivalent of 825 tanker trucks—every day. Men walk back and forth as trucks enter and exit. The trains clank and hiss. DANGER—NO TRESPASSING. ALL VISITORS MUST CHECK IN AT THE OFFICE. RESTRICTED AREA. AUTHORIZED PERSONNEL ONLY. WARNING: YOU ARE UNDER VIDEO SURVEILLANCE. An American flag blows in the breeze.

In the early hours of July 6, 2013, an unattended oil train parked on an incline outside of Lac-Mégantic, Quebec, began to roll backward, eventually reaching sixty mph before it derailed outside of a popular nightclub. The runaway bomb was carrying nearly two million gallons of Bakken crude. The resultant inferno stretched hundreds of feet into the sky. Flames erupted from drains, sewers, and water pipes, torching homes from the inside out. Fire and oil shot half a mile into the river. Manhole covers exploded. Observers felt the heat a mile away. A summer shower most likely saved the town, but even at a quarter-mile's distance, raindrops hitting an open umbrella turned to steam. Forty buildings burned and forty-seven people died, five of whom seem to have been completely vaporized.

To move oil across the country, you essentially have two choices: pipe or rail. While smaller pipelines might gather oil from nearby wells and transport it short distances, there are no major high-volume transmission pipelines servicing the Bakken. Most of the oil leaves on trains. The month I visited North Dakota, the CEO of one major energy-supply company spoke stridently on the advantages of rail over pipeline, citing the flexibility to respond to unpredictable shifts in production and market demand. Addressing the U.S. Energy Information Administration's annual conference, he said, "Despite skepticism, rail may actually be the safest mode of transportation of crude."

Later that fall, both environmentalists and some Bakken CEOs will speak against the proposed Keystone XL pipeline. The former will argue it would be devastating to the ecosystems and public health along its route—and increase our investment in oil. The latter will simply call the pipeline irrelevant. Why risk years of hassle and expense, not to mention tie their own hands in terms of where they can move their product, when a train can deliver oil anywhere in the U.S. today?

From 2008 to 2013, rail transportation of crude increased 45-fold in the U.S. Bakken crude is tanked through many cities, including Chicago, Kansas City, Albany, Cleveland, Buffalo, Oklahoma City, Portland,

Philadelphia, Sacramento, Fort Worth, Minneapolis, Los Angeles, Seattle, and St. Louis. One estimate found more than twenty-five million Americans living within the one-mile "blast zone" of a potential train explosion. Since the Canadian disaster, Bakken oil trains have crashed near Aliceville, Alabama (the amount spilled is unknown, but somewhere near 750,000 gallons; the fire burned for a day and could be seen for ten miles); Casselton, North Dakota (four hundred thousand gallons of oil spilled, twelve hundred people evacuated); and Lynchburg, Virginia (three hundred people evacuated). Trains spilled more oil in 2013 than in the preceding forty years combined.

At the end of 2013, Lynn Helms, North Dakota's top regulator and director of the Department of Mineral Resources, told lawmakers, "Rail has really saved our bacon in this whole business," and announced plans to put together a white paper "to dispel this myth that [Bakken crude] is somehow an explosive, really dangerous thing to have traveling up and down your rail lines."

After the Casselton wreck—some eighteen days later—Helms backtracked, telling reporters he was misunderstood: "Maybe what got to people was the word *myth.* Sometimes they are proven true or false. . . . What we need to do is do the science, collect the facts, and see where that takes us."

Conventional crude oil is hard to burn—and rarely explodes. In January 2014, after witnessing the fires caused by recent derailments, the Pipeline and Hazardous Materials Safety Administration warned that Bakken light-sweet crude might be more flammable than traditional heavy crude. A month later, the *Wall Street Journal* revealed its own analysis, which found Bakken oil boasted a higher vapor pressure and a greater concentration of combustible gasses than samples from other wells worldwide. The following month, the Canadian Transportation Safety Board cited the oil's low flash point (on par with unleaded gasoline). Two months after that, the North Dakota Petroleum Council, an industry trade group, released its own study, finding nothing out of the ordinary.

The most likely culprit: volatile gasses (such as ethane, butane, and propane) that are not being removed at the wellhead, a practice common in other areas of the country where similarly gassy crude is being fracked out of shale. Even in a deregulated, "pro-business" state such as Texas, companies have to strip volatile components from the oil, prompting one Houston industry man to tell *Reuters* this year, "It's a little like the wild west up in the Bakken, where everybody gets to do what they want to do."

Most of the Bakken oil is shipped in the DOT-111 tank car—those ubiquitous long, black barrels originally designed in the 1960s and known for decades to be unsuitable for transporting dangerous material. Less than half an inch thick, with easily punctured ends and poorly protected valves, the cars have been called "soda cans" and "ticking time bombs" (the latter by Senator Charles Schumer from New York). Despite widespread concern—the National Transportation Safety Board has been calling to replace or retrofit them since 1991—more than seventy-eight thousand substandard DOT-111s remain on the rails. Which means seven hundred thousand barrels a day of volatile oil leave the Bakken in outdated, unsafe cars that will travel, on average, some sixteen hundred miles throughout the greater United States.

At the end of July, the U.S. Department of Transportation proposed new guidelines for trains carrying Bakken oil—reduced speed limits, better breaking technology, and a two-year phase-out of DOT-111 cars (unless they're retrofitted to comply with new safety standards). The report included an analysis confirming that Bakken crude "tends to be more volatile and flammable than other crude oils." The proposal was open to public comment for sixty days, during which environmental groups called for stricter regulations—and the DOT-111s to be taken off the rails immediately. The oil lobby advised against "overreacting," calling the two-year timeline "unfeasible" and foretelling widespread economic disruption and a $45.2 billion cost to consumers.

Four months later, a North Dakota state commission passes a standard that—starting in April 2015—requires companies to remove some of the volatile gasses before moving Bakken crude to market. The North Dakota Petroleum Council, along with the largest oil producers, protests the expense as being unnecessary. Environmental critics complain that the process doesn't go far enough—the oil won't be "stabilized," as it regularly is in Texas, but merely treated using existing equipment that isn't as effective. Furthermore, the standards won't specify where the oil must be processed—at the well, the rail yard, or other locations before shipping—and will simply create another problem: a large volume of extracted, highly combustible liquid that must still be shipped by train. Meanwhile, new federal guidelines give the industry until 2025 to replace outdated cars, and even supposedly "safer" non-DOT-111 tankers (containing oil with a lowered vapor pressure) continue to derail—and burn.

The month I'm in North Dakota, the news breaks: the United States has surpassed Saudi Arabia to become the top oil and gas producer in the world. The summer of oil! The long fever dream of energy independence! But this global dominance most likely cannot last. The production of shale wells drops more quickly than conventional wells. To sustain the current million-barrel-a-day output, some analysts say the Bakken would need to drill twenty-five hundred new wells each year. Iraq—with its conventional oil deposits—would only need about sixty, and the wells could break even if prices dipped as low as twenty dollars per barrel. We have to work so much harder to get our oil out of the ground. In the end, there is only one sure thing in the Bakken. At some point, the oil will run out.

I enter the museum. On a table a placard explains:

> What is the origin of oil? The organic material that is the source of most oil has probably been derived from single-celled planktonic (free-floating) plants, such as diatoms and blue-green algae, and single-celled planktonic animals, such as foraminifera, that lived in our ancient past.

In other words, bury organic matter under layer and layer of rock, add pressure, heat, and, say, millions of years, and you might strike it rich.

Above the placard stands a one-sixth model of a *T. Rex.* In the next room, a giant *Stegosaurus* skeleton lies articulated in a bed of sand, as an *Allosaurus* stalks the ledge above. A *Thescelosaurus* lumbers past a downed *Triceratops,* who happens to be named "Larry." The bones of an *Edmontosaurus* slumber peacefully. Another fifteen-hundred-pound *Triceratops* skull sits on a block. Most of us wander Dickinson's dinosaur museum's modest exhibit hall in silence, as requested by a sign near the entrance. The reverential spell is broken when a boy shouts, "Wow, Dad. That was worth the trip!" Meanwhile, a little girl in a *Frozen* T-shirt considers a model of a prowling *Velociraptor.* She fixes her grandmother with a serious look. "Good things these things are extinct."

Here's how remote some of this highway is. Outside of Amidon, south of Dickinson and just beyond the Bakken boom in the southwest corner of the state: two bikers down, sprawled out on the southbound side of the two-lane road, both in leather, one with blood on his face. A third biker grimly walks up and down, picking pieces of metal and plastic off the highway. The sheriff is here, but the accident is recent. I'm the first car stopped in my direction. Slowly, a line grows behind me. After a while, the sheriff waves on traffic, and I inch around the bikers, who still lie face up. Thirty minutes down the road, I pass an ambulance racing to the scene.

Last November, a plastic pipeline broke on the Montana–North Dakota border, spilling seventeen thousand barrels of saltwater that flowed nearly three miles, eventually entering Big Gumbo Creek, south of Marmarth. Soil was removed and the creek flushed with fresh water. (The incident report ends, “Follow up when snow melts.”) In April, a pipe leaked 560 barrels of saltwater and chemicals into the Badlands north of Marmarth. Two months later, a rancher downstream reported he was still waiting to learn from the oil company, Continental Resources, Inc., what was in the water. He had lost several new calves and cows. The spill occurred in the Little Missouri National Grassland, near the boundary of Theodore Roosevelt National Park, which sits atop the Bakken. The grasslands, managed by the Forest Service, are home to more than six hundred wells; flares are visible just beyond park borders. There has been talk of drilling within the park's seventy thousand acres, which inspired Roosevelt's own environmentalism and which many consider “the cradle of conservation.”

To get to Marmarth, skirt the edge of the Little Missouri grasslands, where wildfires and lightning can ignite veins of underground coal that smolder for decades, releasing smoke and fumes that warp the junipers into telltale columns. Head south to Bowman, North Dakota, where herds of bison graze, and follow the train tracks west out of town. The road drops into the Badlands just outside of Marmarth, where the green grass gives way to brown buttes and burnt outcrops and cell service dies. Once you cross the Little Missouri River, you know you're there.

When I arrive at the bunkhouse—at the end of Main Street, next to the tracks and a railroad sign that says MARMARTH, despite the fact that there is no depot and the trains don't stop there anymore—it's empty. In 1907, the president of the Chicago, Milwaukee, and St. Paul Railroad named the budding town after his granddaughter, Margaret Martha. To call Marmarth a ghost town is to do a disservice to the 125 good souls who live there, but the town has lost a lot since 1920, when its population stood at 1,318. Today, empty lots and abandoned buildings line the streets. Beside the Mystic Theater (built 1914) stand two old metal jail cells, one door swinging open. The Barber Auditorium, its pediment stamped 1918, remains boarded up. Painted on the entrance to the Cactus Club: STAY OUT.

Marmarth has always been a place for hunters of some sort. Theodore Roosevelt shot his first grizzly to the west of here and his first buffalo to the north. He was searching for solace after losing both his wife and mother on Valentine's Day 1884. The town experienced an oil boom in 1936, and my WPA guide mentions a nearby well pumping an unusual crude: "apparently high in gasoline and kerosene content, very light, but darker in color, and with a somewhat different odor." Today, there are flares and pump jacks outside of town, but it's not like up north. Five years ago, it was hard to get a room at the bunkhouse—Marmarth was crowded with roughnecks as older wells were overhauled with new secondary and tertiary recovery techniques designed to increase the pressure and flow, but things have calmed down since. That is not to say the town has remained unchanged: its elementary school (prekindergarten through eighth grade divided between three classrooms) will open a $1 million addition this fall, funded in part with a no-interest loan from the county's oil and gas royalties. Eighteen students are expected to start the new year. The nearest high school is twenty miles away in Montana.

On the afternoon I arrived, the single sign of life was a white oil-company truck parked in front of the town's only bar, the Pastime Club & Steakhouse, where a bartender might ask a stranger, "Are you a bone picker?"

Henry Fairfield Osborn, president of the American Museum of Natural History and the man who named *Tyrannosaurus rex*, wrote in 1909: "The hunter of live game, thorough sportsman though he may be, is always bringing live animals nearer to death and extinction, whereas the fossil hunter is always seeking to bring extinct animals back to life."

"Ready to go back in time?" the guy sitting beside me says, rather dramatically. He's from Long Island and is also an amateur. We're in a dusty Suburban pitching itself headfirst down a sharp slope into the Badlands. Through the cracked windshield I see a moonscape eroded out of the prairie: a mottled topography of red, brown, black, yellow, green, and gray studded with naked buttes—the sediments of the sea, silt and clay deposited and then worn down, epochs later, by water and wind. In places, the buttes are scorched and collapsed by burning coal turned into ash. Nonnative sweet yellow clover has choked out the prairie grass that usually grows between the desolate washouts and draws; in parts, the clover stands waist-high. Above, sparse thickets of cottonwoods, maybe a green ash, a few ponderosa pines. Below, baked beaches where alien outcrops of rocks bloom in strangled, man-sized shapes. A landscape of hard eternity, home to rattlers, bull snakes, prairie dogs, pheasants, foxes, coyotes, pronghorns, bobcats, mule deer, minks, and ever-thirsty toads. My companions and I are dressed in paleontologist chic: tan pants, wide-brim hat, long-sleeve button-down, boots, bandannas. As our vehicle lumbers down the hill to the desolate floor, we pass a rock layer known as the Cretaceous/Tertiary (K/T) boundary, a thin line of tan clay beneath a band of coal that pinpoints the "sudden" geological moment when the dinosaurs disappeared.

These aren't the Badlands of South Dakota, which are thirty million years younger and far more popular. These are the Badlands that in 1864 Brigadier General Alfred Sully of the U.S. Cavalry, busy marauding against the Sioux, described as "hell with the fires out." Nearly one hundred years later, John Steinbeck wrote in *Travels with Charley:*

> I was not prepared for the Bad Lands. They deserve this name. They are like the work of an evil child. Such a place

> the Fallen Angels might have built as a spite to Heaven, dry and sharp, desolate and dangerous, and for me filled with foreboding. A sense comes from it that it does not like or welcome humans.

I am part of a team digging up a juvenile *Torosaurus*, a relative of the better-known *Triceratops*. Officially, I am a volunteer member of the Marmarth Research Foundation, a scientific nonprofit dedicated to fossil education, excavation, and curation. Why are we here? For the state's *other* boom. North Dakota in general—and Marmarth in particular—has found itself in the middle of a modern bone rush.

The dig site has been worked for six weeks over three summers. I set my pack where someone previously found a dinosaur skull. There are no clouds and no shade. We're digging through layers of sandstone and clay with chisels and hand picks, looking for signs of any remaining fossils: hard "rocks" with suspicious bony bands or marrow holes, or even just chunks of ironstone that might indicate we're in the right area. *Ting! Ting! Ting!* go our hammers and chisels as we carve small benches into the face of the butte. Hitting bone sounds different—stop scraping then. The ground is littered with hard red-black mineral balls called *concretions* (damned fossil look-alikes!). We find nuggets of bright-orange amber, which one of us collects in a medicine bottle, hoping to turn them into jewelry. We dig into yellow spheres of soft, smelly sulfur. We hit a layer of "veggie matter," and turn up delicate specimens of leaves, mainly *Dryophyllum*. The rock usually cleaves along the plane of the leaf; on one side is an imprint, or image, and on the other is the leaf. (Look for the faint raised ridge of a stem.) We unearth thick mats of spongy organic matter, the veins of the plants perfectly articulated. I hit an ancient tree branch reaching through the rock; the wood is flecked with shiny dark cubes that smudge my fingers black. "What's that?" I ask. Someone replies, "Charcoal, of course," and I think of fossil fuels.

We are excavating what is called the Hell Creek Formation: a bed of sandstone, siltstone, and mudstone deposited sixty-five to sixty-seven million years ago by rivers flowing to the long inland sea that stretched north–south across what is now North America. The formation contains the last of the dinosaurs before they went extinct.

The Bakken is about eighty-five hundred feet beneath the Hell Creek Formation—and some three hundred million years older—but geologists say the oil was formed (cooked, as it were, out of the shale) at about the same time as the dinosaurs roamed and died. That is, both treasures being hunted and hauled out of the ground of North Dakota—the oil and the dinosaurs—are roughly the same age.

We carry our own water here; at least three liters a person. Water is precious, the subject of great conversation. What comes out of the tap in the bunkhouse gives out-of-towners the runs, so we pack big recycled plastic bottles filtered by reverse osmosis. In June, a pipeline spilled two thousand barrels of saltwater eight miles southwest of Marmarth, where the stream of pollution ran for about a mile, before soaking into the earth.

Lunch time: we sit in the holes we've just dug and throw our apple cores right off the hill. Eight hours of chiseling goes by surprisingly fast, despite the unrelenting sun and dust. At the end of the day my back is wrenched and my hands are sore. Tyler, our leader, who is not with us today, found a *Triceratops* right over the next gumdrop butte. We spend forty-five minutes searching but never find his excavation or the trace fossils leading up into the hill. We prospect around the base of the buttes for fallen bone fragments. We sift through the slopes of giant anthills, where the industrious diggers bring up fish scales, teeth, tiny pink vertebrae, and other fossils blocking their tunnels. We are ghouls with good intentions, looking for signs of death. Out here, everything crumbles, falls apart. The Badlands don't preserve—they erode—which is why we find no modern bones, only age-old buried ones. The most recent corpse I come across is a desiccated bull snake. The silver sage smells wonderful when we brush against it, but the sudden whir of a grasshopper might make a nervous prospector jump, were he wary of rattlers. I take a picture of what I think is the K/T boundary on a high butte looming over us—between two gray rocks, an ominous band of coal, marking the lands below as belonging to the dinosaurs.

A dusty blue truck pulls up to the bunkhouse. A thick suntanned guy in his twenties walks in the door. He's been here since April, when he came up from Arkansas. He was hoping to drive a truck—"and not look like this at the end of the day"—but they gave the job to someone with more experience. When we shake, his hands are black with grease and dirt. (A sign taped to the bunkhouse washer says ABSOLUTELY NO!!!!!!! GREASERS IN THE WASHER OR DRYER. —THE MANAGEMENT.) The guy works maintenance for a company that lays concrete all over the state. He spent two weeks working on an intersection outside of Williston, where he lived in a man camp. "I fucking hate Williston," he says. Now he's part of the crew building the new airport outside of Bowman. He's in charge of his own repair truck; it's better than working in Arkansas. "Now, if you'll excuse me, I'm going to go get cleaned up," he says. "I've got some drinking to do."

Nearly every night at 4:00 a.m. the Burlington Northern Santa Fe rolls past the bunkhouse, just feet from our front door. I wake to the train—first the loud, breathy blast of the whistle, then the clacking and creaking of tank cars trundling by.

The Marmarth Research Foundation is the brainchild of paleontologist Tyler Lyson, a thirty-one-year-old Marmarth native and Yale PhD who just finished a postdoctoral fellowship at the Smithsonian Natural History Museum in D.C. and will be taking the post of Curator of Vertebrate Paleontology at the Denver Museum of Nature and Science this fall. The MRF digs on private land, mostly ranchland belonging to Tyler's family. His father, Ranse, is a retired oil production foreman, and when I email Tyler out of the blue expressing my interest in oil and fossils, he invites me to dig, saying, "Sounds like an interesting idea: two very different ways that the past is enriching our state." I'm onboard after signing an agreement not to reveal GPS coordinates or scientific information that might aid "fossil poachers."

Tyler is a dino-hunting prodigy. A restless, bearded live wire who often says "right?" (rhetorically) and can scamper up a butte like a mountain goat, he found his first fossil (the jaw of a duck-billed hadrosaur) at age six and got himself hired as a guide by a professor from Alabama when he was in the fifth grade. He has a particular expertise in turtles, and one of his finds includes a turtle "graveyard" that has yielded more than a hundred shells and three new species. This spring, Tyler was part of a team that announced the discovery of *Anzu wyliei,* a new five-hundred-pound, eleven-foot-long, feathered-but-flightless beaked dino-bird with a crested head and razor claws, something akin to crossing an emu with a reptile, and nicknamed in the press "the Chicken from Hell."

His biggest discovery, however, has been "Dakota," the sixty-seven-million-year-old, four-ton *Edmontosaurus* he found on his uncle's land in 1999 when he was only sixteen. Painstakingly unearthed over the next seven years, "Dakota" revealed itself to be a rare "dino-mummy"—a fossil with nearly all of its bones, ligaments, tendons, skin, and scales preserved. The subject of a TV special and books for both adults and children, "Dakota" has traveled as far as Japan, though currently it is on display at the newly expanded North Dakota Heritage Center in Bismarck, which in a few months agrees to donate three million dollars to the Marmarth Research Foundation for the right to permanently display the fossil—which might explain, in part, that confidentiality agreement.

We're working a tight site. Tom, our field coordinator, says, "There just aren't enough places to get at it." *It* is the four-hundred-pound

Thescelosaurus jacket we're about to flip over. A jacket is the excavated rock around a fossil that has been covered over with foil and then swaths of burlap soaked in plaster. A jacket might contain many bones—here, maybe twenty or so, Tom guesses. For support, we have built a wood frame and plastered it to the top. ("We're all getting plastered in the Badlands!" is a joke I've heard four times today.) Once we've carted the plaster block to the lab, someone will likely prep it—cut it open and painstakingly sift through it—over the winter, which is the equivalent of a rush job. Scientific priorities change over time, and one of the volunteers tells me that some museums have jackets they're still only getting to from the 1800s.

We're caught between the butte and some bones sticking up from the ground we don't want to disturb. A film crew came by earlier and shot about three hours of us chiseling and plastering for a *NOVA* documentary. We waited for them to leave before trying to flip the giant jacket. We contort ourselves into odd positions—all hands on the jacket, feet placed with care—and heave together on a three-count. Once started, you have to keep the block moving—just watch your fingers and toes. A miracle: no earth falls out from the unplastered bottom. We carefully shave a foot or so off the top of the jacket, which then weighs a more manageable 150 pounds. We plaster a thin topcoat, and the next morning we gingerly set the jacket on a burlap stretcher hung between thick poles—like some great beast caught on safari—and then carry it up the hill and onto the truck. We drive back to the lab on a red clinker road built by an oil company that had no luck with its well.

On a Monday night at the Pastime—where a painting of a pump jack hangs above the pool table, the jukebox plays Waylon Jennings alongside Katy Perry, and a sign behind the bar advertises all-you-can-eat crab every Friday from five to nine—the bartender might charge out-of-towners anywhere from fifty cents to a dollar more for a bottle of beer, bringing the total to an even three bucks. A few locals bend over their drinks. The bar can be rough, but not tonight. I've already heard of one legendary fight at the Pastime: a troublemaker brought up the topic of "liberalism" and "open-minded thinking" and the resultant brawl spilled out into the street and down the block into the bunkhouse, which, since that night, has had a lock on its door.

At a high-topped table sits a very tattooed French palynologist, who studies ancient pollen and spores and whose samples will help date the Hell Creek layers. The *NOVA* film crew (a British director, a cameraman, a boom guy) has ordered a pizza from Baker, Montana, which will arrive in an hour. When the director of the Smithsonian Natural History Museum walks in the door and buys me a beer, he tells me he has been coming to town for the past thirty-four years. "You could dig here forever," he says. "It's raining dinosaurs."

The MRF lab has enough jackets stored away to last for years. They rest on huge wooden pallets in the warehouse and line the metal shelves against the walls. Others are secreted away in locations around town. A field coordinator named Stephen picks a rock from an open jacket and points to a nearly invisible line less than a millimeter thick winding its way through the stone. "Look—fossilized skin from a *Thescelosaurus.*" I nod and pretend I see what he sees. The man knows skin; he's one of the ones who prepped "Dakota." This piece was collected at least a decade ago, before Stephen came onboard. The jacket we just hauled out of the Badlands sits on a wooden table. There's a chance Stephen will be taking it home with him to Michigan at some point.

The collection room is a crypt of wonders that diminishes the casual visitor: such impossibly large bones—what giants left them behind? Femurs out of *The Flintstones,* a *Triceratops* horn longer than my arm. I ponder a number of turtles huddled in situ in stone, like some half-buried suicide pact. Barb, ten years at MRF and my guide for the tour, points toward the back: "In these cabinets is arguably the world's best collection of turtles." She's right and she's wrong:

sure, there are drawers and drawers of shells, heads, and tails—not to mention a fully articulated turtle foot that's so beautiful I almost want to cry. (Most turtle species cross the K/T boundary—that is, they didn't die out in the global event that doomed the dinosaurs.) But the cabinets contain countless other finds: fearsome claws from bird-beasts, delicate reed-like tendons, sly and sinister dinosaur beaks, a massive tail that was smashed in battle and then healed, the vertebrae fused.

When I leave the lab that night, another preparator also named Stephen is bent over a magnifying glass, picking away at a massive *Triceratops* vertebra with a miniature pneumatic jackhammer called an air scribe. He's been working on this bone for three years. He says, "It's beautiful, isn't it? These parts here"—he holds up the swooping sides—"rise up like wings." In the next room, Barb is running the air abrader, her hands in a big glove box—like a scientist fighting an infectious disease—as she blasts a giant bone with baking soda. The fossil looks beautiful, as if settled in snow. Giant air compressors whir as she picks away the past, one grain at a time.

On my way out, I stop by the tabletop sandbox, which is filled with red garnet sand and used as a soft space to puzzle out how fossils fit together. Two big pieces of a hadrosaur tibia stand upright, as if growing out of the sand, while next to them someone has carefully laid out a handful of turtle fragments. There is something unexpectedly touching about the twenty-odd chips of shell arrayed in an incomplete oval, the central mound of red sand ready—ever hopeful—to cradle the bones.

Later, I will ask Tyler, "Why turtles?" He tells me they have a rich fossil record—they're as common as leaves. He's captivated by the evolution of their bodies, how they lock their ribs into a shell ("It's one of those features that appears exactly once in earth's rich history, like the bird feather"). But what stays with me is this: "You find dinosaurs here and there, but turtles are pretty continuous through time."

A cartoon on the lab's bulletin board shows the "Causes of Mass Extinctions." The Cretaceous period is ended by a meteor, the Pleistocene by a change in climate, and the Age of Man by an "act of stupidity": a caveman bangs a hammer against a bomb that seems to be resting on oil drums. On the bomb is written: GOD IS ON OUR SIDE.

"Ebola: What You Need to Know"; "Gaza Crisis Brings 9/11 Flashbacks"; "Surgeon General Calls for Action to Reduce Skin Cancer Rate"; "Suicide Bomber from U.S. Returned Home Before the Attack"; "Race to Find India Landslide Missing"; "NATO 'Unprepared for Russian Threat'"; "Six Dead in Nigeria College Blast"; "Drug-Resistant Malaria Widespread"; "Why Do Americans Not Buy Diesels?"; "Midwestern Waters Are Full of Bee-Killing Pesticides"; "White House: Ignoring Climate Change Will Cost America Billions"; "U.S. Oil Exports Ready to Sail"; "Should America Keep Its Aging Nuclear Missiles?"; and "Winds of War, Again" are some of the headlines I read late one night while in Marmarth.

In some ways, this moment feels particular, transitory, just a sliver in time. Though it seems unlikely (if not impossible) to me, many interested parties believe impregnable pipelines can be built, water will be recycled, flaring curbed, the earth protected, and safe and spacious cities will arise on the plains. But something about this summer feels endless, eternal. The question is not so much *How did this happen?* but *Why do we always find ourselves here?*

And then, while I'm in Marmarth, the opposite thought occurs. *This is it. We've reached the brink, we had our chance; our turn is over.*

On a dig, one might talk about Victorian pianos, brewing beer, maple syrup, Cretaceous vegetation, or modern sand channels, but most often the work is done in silence—save for the *tink* of hammers and chisels and the incessant scraping of awls and knives. The mind veers from the micro to the macro; if you're not careful, you might get whiplash. You're chiseling away with a short knife at a few inches of tough, barren cliff all morning when out pops a piece of amber and suddenly you remember you've got your nose pressed up against a window looking sixty-seven million years into the past. According to the female volunteers—some of whom prep fossils for this country's major museums—dino-digging remains something of a boys' club. But if paleontology might be conducted with a certain macho swagger, hunting for fossils is fundamentally humbling: again and again, you end up feeling so *small.* Where do I stand on this colossal, shifting earth? As alien as this landscape seems to me today, what sends my mind spinning is to think of it *then*—flat, warm, and wet, lush with ferns and vegetation. A humid coastal plain. Pretty much the opposite of the Badlands around me.

And these bones! Not only do they dwarf us, but these creatures lived so long ago. How do we measure ourselves—and our species—in geological time? A half blink, at best. Once again—why are we here? What could we ever contribute to science, that incomplete story we tell about ourselves? Paleontologists dwell in uncertainty and failure; they're always saying, "We just don't know" and "In fifty years, we might understand a little more." These few feet of excavation, a couple of centimeters of carefully prepped bone. To dig is to be constantly confronted with your own pitiful insignificance.

We move to a possible hadrosaur site just below a ridge with a 360-degree view of the Badlands. Is there a duck-billed herbivore buried in this cliff—or enough of it, anyway, to make our work worthwhile? Within two minutes of brushing soft dirt, I find a reddish-orange rock. I put it on my tongue, and when it sticks, I know: I'm French-kissing a fossil! (The porousness of the bone clings to the surface of the tongue like an old, hard sponge. The sound the rock makes as you pull it away—like softly peeling an orange—is a deep, visceral thrill.) Most likely the piece eroded and tumbled down from a larger chunk of bone, and so with a scalpel and brush I chase the dinosaur into the hill. But in the end it eludes me. We find only broken bits of tendon or perhaps a small rib. I put my piece on a short pile of bone, and we pack our gear and climb back up to the truck.

After a week of digging in the field, I find myself walking through town, unconsciously scanning the side of the road for fossils, studying the bigger rocks and imagining just where I would put my chisel to break open their secrets. At night, the Milky Way careens brilliantly overhead, looking like a band of stubborn sediment in the sky.

On my last day, we go prospecting in an area no one has searched in a decade. We agree to meet at a distant butte around lunchtime. On the ground we find hunks of frill—the wide bony plates that flared out the back of ceratopsian heads—plus scattered sections of vertebrae. I pick up a nugget. "Chunkasaurus," says Antoine, the French palynologist. "A big, exploded dinosaur." Later, I will see a naked rib sticking right out of a hill. I will pick up a square of crocodile scute, or fossilized armor, plus a shard of turtle shell and two small, broken dinosaur bones. One of us will find a stone Indian knife, a long right-handed blade that fits perfectly in your palm with grooves for your fingers. Someone else comes across a coprolite—a fossilized piece of shit. We will walk some eight miles in one-hundred-degree heat, marking the GPS coordinates of anything worthwhile.

Not long after splitting up, I climb to the top of a tall butte, where I run into Antoine again. Antoine lives in Paris. He carries a shovel to dig trenches from which he takes careful samples—he then dissolves the rock layers in acid and studies the organic remnants under a microscope. The ancient spores and pollen help date and describe the environment. At breakfast this morning someone called Antoine "an extinction guy," meaning he studies the K/T boundary, when the dinosaurs died off.

Today he has been searching for the boundary layer, but, alas, no luck. He scrapes at the cliff, frowns, and says, "The story here is complicated." He just came across a rattler in its hole, so we double back around the butte. We scan the horizon, but there is no sign of our group. They're crawling somewhere in the craters below us. We hunker down in a sliver of shade to eat lunch by ourselves, and I ask Antoine about his work. He tells me a lot of scientists study the late Cretaceous right up until the extinction. It's popular: there are dinosaurs, drama, and death. But he's curious how life clawed its way back from the devastation. He takes a bite of his sandwich and tells me, "I think that is where the interesting story lies."

The K/T extinction event—now more properly known as the Cretaceous–Paleogene (K/Pg) event, though the more recent name hasn't really stuck—wiped out all non-avian dinosaurs and ended the Mesozoic Era. When I was growing up, what killed the dinosaurs was a subject of great debate. I remember picture books filled with hypotheses (volcanoes, continental shifts, natural selection, and so on). Science has since solved that mystery in spectacular fashion.

Some sixty-six million years ago, a six-mile-wide asteroid—speeding in low from the southeast around forty-five thousand miles per hour—crashed into the Gulf of Mexico right off the town of Chicxulub on the Yucatán Peninsula, sending up unimaginable amounts of flaming dust and debris that scorched the surface of the earth before darkening the skies and bringing about an "impact winter"—in essence, causing a change in atmospheric composition that led to a change in climate that led to trouble for those on earth. Firestorms. Earthquakes. Mega-tsunamis. Shifts in ocean chemistry. Mass extinctions as years of darkness and cold gave way—as the particles settled out of the atmosphere, many falling as acid rain—to centuries of intense global warming (from the CO_2 blasted into the air). The event is at once spectacular and familiar. Three-quarters of the species on earth eventually died in the fallout.

You can change a climate from without, say via flaming asteroid, or you can change it from within, say via the steady combustion of fossil fuels. If getting oil out of the ground is hard on the earth, burning it is worse. That is, one way to consider the Bakken is as a site of local hemorrhaging, the symptom and symbol of a greater, far more devastating disease. Since the Industrial Revolution, humans have pumped 365 billion metric tons of carbon into the air by burning oil, coal, and gas. Atmospheric CO_2 is up 40 percent, and we are on track to double the preindustrial level in the next thirty-five years. (Meanwhile, concentrations of methane have already doubled.) As we know, greenhouse gasses warm the planet. By 2050, temperatures might rise as much as seven degrees. As Elizabeth Kolbert writes in her book *The Sixth Extinction*, "This will, in turn, trigger a variety of world-altering events, including the disappearance of most remaining glaciers, the inundation of low-lying islands and coastal cities, and the melting of the Arctic ice cap." She also points out that the oceans absorb a fair amount of this CO_2, which makes them more acidic, or poisonous to life.

Geologists employed by Pemex, a Mexican oil company, originally discovered the Chicxulub impact crater. Already an ever-growing fleet of ships is fracking the ocean. The wastewater is treated and—supposedly harmless—dumped into the sea. "Deep Water Fracking Next Frontier for Offshore Drilling" is a headline *Reuters* runs while I'm in North Dakota. According to the article, the "big play" is a formation called the Lower Tertiary in the Gulf of Mexico, which just happens to be the layer of rock on top of that earlier, more fateful big boom.

In a few months, I will cross the Gulf of Mexico, leaving from Texas and passing deepwater drilling platforms before sailing near, but not over, the Chicxulub crater. I will stare out at the ocean—dark blue to the horizon in every direction, the black depths below—and I will shudder to think that our clumsy actions have the power to raise the temperature, by even a single degree, of these endless waters that bathe us.

The K/T event was not the planet's most cataclysmic upheaval. That distinction goes to the far older Permian–Triassic extinction, a period 252 million years ago also known as "the Great Dying," when 90 percent of all species on the planet died out. It's the only time the insects were killed, too. The planet grew incredibly hot; sea temperatures might have risen as much as eighteen degrees. The world nearly wiped itself clean. The exact feedback mechanism is still being debated—Siberian volcanism, global warming, ocean acidification, the melting of frozen methane, the production of hydrogen sulfide—but most scientists think the dying began with the release of greenhouse gasses.

The catastrophic volcanism that many believe triggered the Great Dying is thought to have released less carbon annually than we humans currently emit into the air.

Convert the heat our emissions add to our planet into something more visibly spectacular: imagine four hundred thousand of the atomic bombs that scorched Hiroshima detonating every single day.

Today, long-frozen methane in the Siberian permafrost is warming—and then exploding. Formerly trapped gas is also being released along the coast, where methane plumes bubble to the surface of the sea (in concentrations ten to fifty times greater than normal). Methane is a far more powerful greenhouse driver than carbon dioxide.

When I was in North Dakota, the internet was abuzz about a mysterious giant crater that suddenly appeared on Russia's Yamal Peninsula. The crater, which was spotted from a helicopter by oil and gas workers, was thought to have been formed by the ejection of underground methane melted out of the permafrost. The past two summers on the peninsula had been unusually hot. More large craters will continue to be found. *Yamal* translates into "end of the world."

When I think of extinction, I picture something gradual. But that isn't how it happened for the dinosaurs—and it's not what's going on today, speaking in geological time. The day before I left for North Dakota, a study titled "Defaunation in the Anthropocene" came out in the journal *Science*. The gist: we are in the midst of the earth's sixth mass extinction. The culprits: do I even need to say? The study points out that since 1500, a total of 322 species of terrestrial vertebrates have become extinct. Those that have survived our ever-growing presence have declined 25 percent in population. Earlier that spring, a Duke study revealed that plants and animals were going extinct at a thousand times the rate they were before humans appeared. Some twenty thousand species currently hang in the brink.

The pace seems to be quickening. A month after I returned from North Dakota, the London Zoological Society reported that over the past four decades the world's wildlife population had been cut in half. Rivers have emerged as a particular killing zone—freshwater animals are down 76 percent. Meanwhile, turtles—which handily survived the K/T event—have been reduced by 80 percent. In about the same span, the human population has doubled.

The larger animals aren't repopulating; smaller animals are taking over, a pattern borne out by other extinction events. Today that means rodents, who help spread disease. Wheels within wheels. All I can think of is that we should have known it was coming—the meek shall inherit the earth.

Of course, the Bakken boom will slow—if not go bust, at least partly—as oil prices fall from the summer of 2014 to the beginning of 2016. During the downturn, Saudi Arabia keeps its production high—glutting the market—in what most consider to be an effort to cripple U.S. shale production. In the Bakken, small operators get squeezed out; the number of active drilling rigs plummets. Oil-field fatalities rise, perhaps due to companies cutting corners. North Dakota no longer boasts the lowest unemployment rate. The governor calls for budget cuts; the state announces it will no longer offer free vaccines to children. (The Legacy Fund cannot be tapped until 2017.) Many newcomers leave, if they're able. Rents drop. In Williston, Walmart lowers its hourly wages, while Heartbreakers, the town's last strip club, closes. (It reopens as a gay bar.) Downtown storefronts and apartments stand empty. A developer tells *Reuters*, "It seems like people are on the fence, waiting."

Nevertheless, the U.S. remains the top oil and natural-gas producer in the world, as drillers prove more flexible than imagined, cutting costs and pumping more oil to stay afloat. Production in the Bakken never falls below a million barrels a day. By mid-2016, the price of oil crawls back up to $50 a barrel. An industry analyst tells the *Wall Street Journal* that that price should signal an industry-wide "all clear," as companies weigh whether to complete more wells. On the campaign trail in Bismarck, Republican presidential nominee Donald Trump promises to slash environmental restrictions and unfetter the fossil-fuel industry. The candidate has called global warming "bullshit."

Meanwhile, climate scientists warn the tipping point is closer than we think. Energy production in the U.S. is leaking far more methane than previously calculated—potentially offsetting recent reductions in carbon-dioxide emissions. The year 2014 is the hottest ever recorded, until 2015 breaks the mark by the largest margin yet.

My last night in Marmarth and we're racing to Mud Buttes to beat the dying of the light. A stretch of Badlands just out of town, Mud Buttes is one of the best places in the world to find the primary markers of the K/T boundary, to put your finger on the moment of impact. We speed down a gravel road as Tyler lists off the physical evidence, debris thrown from the crater that shows up in rock layers worldwide: iridium (a dense metal rare on earth but common in asteroids), spherules (glassy beads of once-molten rock, similar to what I saw at the Trinity site in New Mexico), and "shocked" quartz (the mineral's crystalline planes deformed under great pressure—a phenomenon first noted after nuclear testing).

The sun sinks lower, turning a mushroom-shaped cloud on the horizon a stunning pink and lending our errand an even greater sense of apocalypse. But the mood in the truck is buoyant; some are drinking beer, excited to be nearing the end of something. A fingernail moon hangs in the darkening sky. Flares burn in the distance; others snarl in pits by the side of the road. We jump out of the car and scamper across the Badlands, heading for the butte. As he runs, Tyler calls back, "This ecosystem is thirty to forty thousand years before the rock falls."

Later, Tyler will say: "You want me to tell you about the survivors? After the rock hits, it sends out a thermal pulse, so at least a good chunk of North America gets fried as things come flying back in from space. Everything gets sent up into the atmosphere. Months up to maybe a year of varying degrees of darkness. The cold—that's what I think largely kills the plants. And anything that relies primarily on plants is shit out of luck. Anything that's big—over a meter, roughly—goes extinct. So that pretty much takes care of all dinosaurs. The largest land turtle. The largest alligator. The big things in the ocean."

Tyler squats before one of several trenches dug out of the butte. He points to a lighter stripe of red-tinged rock beneath a band of coal and says, "This is the actual boundary, right here." Antoine talks us through the layers: the spherules, which landed first, because they were heavier; then the "shocked quartz," followed by the iridium anomaly. I run my finger along the boundary, rich with cosmic dust and melted chunks of Mexico that rained down on North Dakota. A line not many fossils cross, those few centimeters marking the end of an era. Tyler reaches in. "Age of dinosaurs," he says, then raises his finger a fraction: "Age of mammals."

"If you have a slow metabolism, you're better able to survive. If you're small, you're better able to survive. And if you're living in the water, or burrowing, you're better able to survive—things like crocodiles, some lizards, and turtles. A lot of the big animals get taken out. But then they

are rapidly replaced. It opens up all these niches. And mammals very, very quickly take over."

Some of us walk to the edge of the butte to watch the sun set. But Antoine's not finished. He moves his hand up the rock face. Annihilation: the planet's been scrubbed free of animals—not many fossils here. Then comes a sharp increase in fossilized fern spores—known as the "fern spike"—as these plants were among the first to recolonize the seared landscape. Higher up, he wrestles a chunk out of the butte. At last, a sign. "Look. Do you see what those are?" I do not. Brown coils wind through the black-flecked rock like thick strands of spaghetti. After a pause, he tells me: "Worm burrows."

My mind tunnels inward. I think of family and friends back home, lives left in the rearview. I think of rigs going up and men breaking down: in stairwells, in parking lots, in temporary trailers. I think of those who know the land and those who just rent it. I think of fires, water, and the hard, cracked earth—and the part of us that longs to break it wide open. I think of what this looks like from space, or from the other end of eternity. I think of things buried, the bones left behind. I think of long sacrifices and short-term gains. And, finally, I think of my daughter, her love of dinosaurs, how I haven't spoken to her or my wife all week except over email and text, since I was beyond cell service, beyond a lot of things, really. How will I tell them what I saw in North Dakota?

The air grows cooler. A few lowing cattle climb the butte above. From somewhere unseen, a bird chirps twice, then stills—the song of a survivor, a distant dinosaur relative. Soon, we'll be gone, but for now we linger on the edge as silence falls across the Badlands, and the sun sets on this world, and the last one, and the one that is to come.

2014–2016

Three Minutes to Midnight

What do you do when you've exhausted the past and the present—when all those looks *back* and *at* tell you nothing about what lies *ahead?* History's a meddler, forever hammering at your door, unannounced, just when you should have expected it but didn't. What if, as the years pile up, you realize there might be nothing salvageable—nothing you can learn, anyway? Present life seems a muddle, at best, and yet something tells you to keep going—there must be an answer somewhere down the road.

It turns out you would do what always has been done in this country—you would go west, light out for the territories, follow the setting sun to the promised land, Shangri-la, the golden coast, the endless Eden, the dream incarnate, specifically, in my case tonight, Los Angeles, La-La Land, twenty minutes ahead of the rest of America, ever on the cusp—of the Next Big Thing, of finding nirvana, of falling into the sea.

Want to see the future? The old clichés still apply to California. But here's the secret, the final twist: you go west, young man, and then you go down, down, down. . . .

The man's voice on the phone is deep, conspiratorial: *"I'm going to scare you."* He uses certain refrains: "some people believe," "my sources say," and "people tell me," as a way of keeping himself at one degree of remove—from certainty, from absolute prophecy, from the lunatic fringe. After our conversation, he will email me some links, saying, "You will need to determine their veracity, while keeping an open mind." He's no chump—he doesn't believe *everything*—but he keeps his ears open, and this summer his sources are chattering like crazy. The picture is grim, the dots are connecting, there's just too much to ignore. An overwhelming confluence of events points to something BIG on the horizon—starting in July and culminating in September—that will be bad news on a planetary scale. And here's what Robert Vicino's people are telling him *right now,* delivered to me, the skeptic, in a looping, oddly charismatic monologue that runs more or less uninterrupted—try as I may—for the next hour and a half.

Have you heard about Jade Helm? It's a military exercise taking place across America. Starts in July, ends in September. It's nationwide; they're using major *equipment. It's believed by my sources that the government is going to be enacting martial law. Why this summer? Because of the recent racial tensions? Nah. Something is coming and they can no longer keep it secret. There are several options:*

One, they're finally going to pull the plug on the economy. We all know it's based on fiat, a big Ponzi scheme. Economic collapse is coming. They're going to do a reset, close the banks, which means you can't get money, you can't get gas, et cetera.

Two, have you heard of the Shemittah, the ancient Jewish Sabbath year? Every seven years debts are forgiven—and these years often line up with horrible events. The 1933 stock market crash, the one again in 1987. I think 9/11 was a Shemittah year. This is a Shemittah year. It happens in September.

Three, we've had a series of consecutive blood moons. The fourth and final one will occur in September, during Rosh Hashanah—also during a Shemittah year. A lunar eclipse! The Bible says something *is going to happen.*

Four, many believe on September 23 a comet or asteroid is going to hit the planet. Why then? Because at a press conference with Secretary of State John Kerry, a French minister said, "Everybody knows we have five hundred days left before a major, catastrophic environmental change on the planet." And everyone in the room just nodded along! What's five hundred days from him saying that? September 23! They know something. When Janet Napolitano, the Secretary of Homeland Security, left office, she said her successor would have to deal with the largest catastrophe in the history of mankind. They know!

Five, across the country, Walmart and Target have been shutting down stores; hundreds of them are boarded up, fenced in. They fired everybody and now there's high security. The stores are right over the hubs of the government's underground railroad. They're depots. And all are ready in time for Jade Helm. They're part of the drill. Or maybe they're to be used by FEMA.

Now, I don't need to go into the Walmart-NSA connection, do I?

Robert Vicino is the founder of the Vivos Group, a California company that promises "life assurance for a dangerous world" by building a network of secret underground bunkers meant to withstand the apocalypse, and then selling berths in them to the average citizen. I am interested in a shelter Vicino reportedly was readying in the Mojave Desert outside of L.A. Many of the bunkers remain to be built, but they're said to be retrofitted Cold War installations—often hardened telecommunication centers—and it seems ironic that complexes that were once meant to ensure connection in moments of catastrophe are being repurposed to separate the pessimistically prepared from the soon-to-be-extinct. Vivos says the shelters are impervious to flooding, fire, earthquakes, extreme hail and winds, chemical and biological attacks, armed assaults, solar flares, and twenty-megaton nuclear blasts. They have surface defenses, radios, generators, fuel, water, medical supplies, air filters, room for pets, and a year's worth of food for every person, plus resources to grow more underground.

Would-be members apply online. From the applications that come in, Vivos curates a community, the optimal mix of doctors, plumbers, electricians, mechanics, carpenters, cooks, educators, security personnel, and the like. The shelters have libraries stocked with books and movies; the company is assembling a DNA vault. The media director tells me, "It's not just about surviving—but continuing." She also assures me the shelters are surprisingly comfortable. (Citing the plush mattresses, pleasant color schemes, and calming art on the walls—"no violence, no blood"—Vicino calls the accommodations four-star: "not Motel Six, but La Quinta.") The media director says she has lived in the shelters for months at a time and never missed the outdoors. And—with an office overlooking the ocean—she's someone who likes a view.

In Los Angeles, the morning TV news shows the ongoing efforts to clean up the Santa Barbara oil spill, two hours up the coast, where last month a Texas-based underground pipeline failed, spewing 101,000 gallons of crude, much of which flowed beneath the 101 freeway and into the Pacific. Black sludge still seeps into the sea. Cleanup is costing $3 million per day and stretches for more than ninety-six miles. Blobs of tar have washed up on beaches in L.A. The oil company is requesting permission to transport the oil—in as many as eight tanker trucks an hour, twenty-four hours a day—up the coast.

Meanwhile, the newspaper covers the ongoing investigation into a confrontation last year that ended with a black, mentally ill young man shot dead by police only two days after Michael Brown was gunned down in Ferguson. In other news, there were two shootings in south L.A. yesterday afternoon, video of a road-rage fistfight in Hollywood has gone viral, and statewide pollution is putting communities at risk. On the international front, an outbreak of Middle East Respiratory Syndrome has hit South Korea, and presidential hopeful Jeb Bush is calling Putin a "bully," as the U.S. leads military exercises near the Russian border.

Susan Sontag writes, "Apocalypse is now a long-running serial: not 'Apocalypse Now' but 'Apocalypse from Now On.' Apocalypse has become an event that is happening and not happening." Essayist Charles D'Ambrosio adds, "Every fundamentalism focuses on end times, and Armageddon is, in a sense, a rhetorical trope, an emphatic and overwhelming conclusion, meant to wrap up and make tidy the mistaken wanderings of history."

Some say the Cold War is over. Shelters are a waste of money. But the Cold War is heating up. And, hey, forget the Cold War—what about the holy war? Divisive religions are destroying mankind. You know, we have a superior court judge as a member of Vivos. What do you think of that? A judge is afraid!

These people are smart enough to read the signs. It's the people who don't who are ignorant. They don't have a clue. They don't want to prepare until it's too late. I have disdain for the guy who shows up at the moment of truth knocking on our shelter door, waving his black AmEx. On that day, when everything hits the fan, WE DON'T TAKE AMEX! *We might take your gold, food, guns, ammo, or skills—but that's assuming you can even find the door to knock on!*

At dinner in the trendy Silver Lake neighborhood (named for one of the city's first water commissioners, who built a reservoir there in the early 1900s), I ask two old friends where they would go at the end of the world—when the big one hits, whatever that means. Are they prepared? One friend answers immediately—of course his family has a plan. He and his wife have two meeting spots, A and B, walkable from their house and their kids' school. (They would shelter in place or, if things got really hairy, they would head out of town to a coworker's ranch.) My friend and his wife keep an earthquake kit at home—food, sleeping bags, flashlights, radio, etc.—plus one in the car. The city suggests being prepared to go at least seven days without any assistance. My friend's wife regularly replaces the bottles of water.

Water, of course, is what everyone is talking about. The drought is now in its fourth year. For the first time in California history, the governor has imposed mandatory statewide water restrictions. Reservoirs are empty; groundwater is being depleted (by cities, by farmers, by fracking). As the water is pumped out of the earth, swaths of central California are sinking, some as much as a foot a year. (Picture buried pipes poking through the surface, roads and canals cracking, bridges slipping underwater.) On the U.S. Drought Monitor map, the state is bathed blood red—"exceptional drought." Freeway signs urge residents to reduce usage. Surveying the reservoirs from space, NASA warns only a year's worth of water remains.

Everywhere Angelenos see proof that the desert is seeking to reclaim the city. A conservation sign in a desiccated lawn (now turned to dirt) might declare: H2—NO! As the crooked politician in the classic L.A. noir *Chinatown* reminds us, "Beneath every street there's a desert. Without water the dust will rise up and cover us as though we'd never existed."

At dinner, we nod at the thoroughness of my friend's plan. We sit in a short, respectful silence. Then my other friend says, "When the earthquake hits, I'm diving into the crack. I mean, I'm going headfirst into the magma." He is no less sure of his plan. "Why would I want to be around for the aftermath?"

On January 22 of this year, the *Bulletin of the Atomic Scientists* advanced the hands of its Doomsday Clock—their measure of how close we stand to the eve of our destruction—to only three minutes to midnight. The hands hadn't moved in years. The culprit: our failure to keep climate change and nuclear arsenals in check. The closest we've come to the zero hour was 11:58 in 1953, after the U.S. and Russia both tested hydrogen bombs. In 1991, a few years after the end of the Cold War, the hands stood an optimistic seventeen minutes to midnight. Ever since, we've been inching back toward doom.

Folk singer Phil Ochs: "The final story, the final chapter of western man, I believe, lies in Los Angeles."

What can you say that has not already been said about a city that holds a regular End of the World Party? L.A. is a town with a hair trigger, poised to fend off—or, failing that, embrace—its own destruction. Take, for instance, the "Battle of Los Angeles," when, on the blacked-out night of February 24, 1942, some fourteen hundred antiaircraft shells were fired in an hour-long barrage at a Japanese plane, a wayward weather balloon, or a UFO, depending on whom you ask—a shelling that resulted in a number of civilian deaths but not one enemy casualty.

Paranoia has slipped into certainty. Last summer, the mayor of L.A.—citing the fact that his city is menaced by thirteen of the sixteen federally recognized natural disasters—announced plans to appoint a "chief resilience officer," who would ready the city for catastrophe. The mayor insisted, "It's not a question of 'if,' it's a question of 'when.'"

When it can no longer be kept a secret—what then? Say it's a solar ray, EMP, asteroid, the government enacting martial law, economic collapse, nuclear winter, Yellowstone starting to go (it is, you know—burping and rising and leaking helium-4), or Planet X (the tenth planet in our solar system, which appears in ancient Sumerian stone carvings, caused Noah's flood, and comes around every thirty-six hundred years to wreak absolute havoc)—would the government tell you if disaster was on the way and there was no solution? That would only create panic! Social breakdown and chaos! As good custodians of society, they can't *tell us. It's not what they say—it's what they* don't *say.*

Vivos won't tell me how many shelters they have, or where they're located. They say the press has burned them before. According to Vicino, he had to scuttle plans for that bunker outside L.A.—the one I wanted to see—when a newspaper reporter revealed the location. (The site—a nuke-proof communications center built in 1965—was vandalized twice during escrow, he says.) A bunker is no good if everyone knows where it is. Plans for shelters in Nebraska and Kansas have fallen through, too.

And here my conversations with the Vivos Group come to an impasse. Despite having appeared on a number of news outlets, they have grown wary of reporters. I become a little wary of them. To even get this far, I had to engage in a bit of cat-and-mouse: pre-interviews and the like. A friendly paranoia seems to be the rule. For security reasons, they say I can't be shown a bunker; they must protect the privacy of their owners and the integrity of their sites. The last TV crew to film a shelter—in Indiana—had their phones and GPS confiscated before being driven to the location in a blacked-out van. I agree to such procedures. I promise confidentiality, vagueness, details withheld. No dice. Vicino says, "If your story became a book, then I would take you there."

I'm looking for a good writer to do the story of Vivos: the good, the bad, the indifferent. The stories of the people who've applied—the appeals from mothers, begging for their children. I have thousands of applications asking for financing. They can't even pay $500. A year's worth of food alone is $5,000 per person! I haven't made a dime. I've invested millions! People say I'm capitalizing on people's fears. But look at the major media! They're the ones doing that.

This hasn't been fun. It's been extremely emotionally troubling for me and my family. I've been attacked. People say I'm a scammer, taking advantage of people. There's a book here, a movie. The subtitle would be, "How I Tried to Save the World and Couldn't."

A billboard on Sunset Boulevard advertises the latest *Terminator* movie—NEW MISSION. NEW FATE.—but judging from the flaming city in the background, the future seems the same as ever. Los Angeles is a favorite setting for Hollywood's apocalyptic vision—a sign of narcissism and self-loathing, perhaps, but also proof of the desire to destroy what we love. *Blade Runner*, the *Terminator* series, *Escape from L.A., Volcano, Earthquake, Miracle Mile, Battle: Los Angeles, This Is the End, Resident Evil: Afterlife, Demolition Man, The Omega Man*, and so on. The city has been annihilated so many times: *The War of the Worlds, Night of the Comet, Independence Day, The Core, The Day after Tomorrow, Transformers, 2012, Zombieland*, and most recently this summer in *San Andreas*, a bit of disaster porn starring The Rock—the posters for which proclaim WE ALWAYS KNEW THIS DAY WOULD COME. Remember, this is the city that not only created a postapocalyptic time-traveling cyborg-slash-harbinger of doom, but elected him governor.

Vicino tells me the internet is an ugly place. There are rumors online. One day, I found a site calling Vivos a fraud. It posted what it claimed was a federal lawsuit filed by an owner who had paid over $140,000—and never saw a finished bunker. A week later, the site disappeared.

I email Vicino. In a long reply, he groups the naysayers into seven categories (competitors, disgruntled ex-employees, those who can't afford Vivos, those who don't believe the end is near, those who think preparation is against the Bible, jealous survivalists who are irked that Vivos requires no extreme training, and government trolls). He refutes a number of the specific critiques made on one of the websites—confusion over calculating food in tons versus cubic feet, etc.—and suggests the government created the site to quell a nervous public. He says Vivos isn't cashing in on fear—it's offering a solution. He writes, "The only fraud is the one being perpetrated by these deviants," and points to the many mainstream media outlets that have covered Vivos. He says more than fifty thousand people have applied for space in the shelters—those selected are grateful and often surprised by what they get for their money. He concludes, "The bigger question that should be asked is where will all of these people go when the S.H.T.F.! They will be the ones banging on the other side of the blast door!"

Frank Lloyd Wright: "Tip the world over on its side and everything loose will land in Los Angeles."

A beat-up white van goes clunking down Los Feliz Boulevard without looking back—both its rearview mirrors are totally busted—but this particular vehicle is focused on the future, specifically the end times, which started two years ago and will unfold according to a timeline that entirely covers the back windows and doors in careful, obsessive scrawl. I sit behind the van at a red light and read: Armageddon will involve a galactic federation, the Mayan high priesthood, an archangel, aliens in spaceships, and "the return of magical beings and the guru masters of spiritual knowledge." The first seals are already open, and the van predicts things soon will get ugly, a chaotic acceleration that results in "billions killed and drowned." The year 2031 is the last time we will see the sun and moon. The prophecy ends simply: "good luck."

You say, "But Robert, this is conspiracy stuff." What is conspiracy? My definition: something that's not yet been admitted to by the government. History proves the conspiracy is right. We're all frogs in boiling water—that's how society changes. I know I'm sounding like some far-right-wing religious conspiracy nut, but I'm not. I'm rational. So you believe in God. How can you be so sure you're on the list? That's so arrogant! There's danger all around us.

Ice T: "Los Angeles is a microcosm of the United States. If L.A. falls, the country falls."

Some cities seem forever caught in the crosshairs, the usual suspects for disaster scenarios: L.A., New York, London, Paris, Jerusalem,

Tehran, Beijing, Mumbai, Moscow, D.C. When my wife and I moved our family to St. Louis, I considered us to be safe in the heartland, until I found out we were situated on the New Madrid fault zone, the site, in 1812, of the largest earthquake recorded in the U.S. The massive tremor devastated Missouri, Arkansas, Kentucky, and Tennessee; it woke people in Pittsburgh and rang bells in Boston. The Mississippi River ran backward. Fields became swamps. The earth opened up and swallowed people whole. Today we must all must live with localized apocalypse.

In early 1980-ish, I had a vision that something was going to happen. It was more like an inspiration. I wasn't very religious—I was a lapsed Catholic—and I wasn't very spiritual. But I got the message that something was coming my way, and I needed to build a shelter for a thousand people underground. At the time, I had my own company, a Rolls-Royce, an airplane—I was doing very well. I looked for a gold mine that I could go and harden, but people said I was crazy, so I tucked the idea away. In 2007, after about thirty years, I had a gut feeling that time was almost out. I mentioned it again to my staff, and they said—have you been reading the news?

Joan Didion: "California is a place in which a boom mentality and a sense of Chekhovian loss meet in uneasy suspension; in which the mind is troubled by some buried but ineradicable suspicion that things had better work here, because here, beneath that immense bleached sky, is where we run out of continent."

Vivos sends me a brochure for a luxury shelter to be built in Germany—an overhaul of a massive Cold War bunker that the Soviets dug out of a mountain. "Europa One" already includes a water treatment plant and a train depot; there is space for a zoo, a "Hall of Records," a seed bank, and a vault for the world's treasures. The brochure calls it a "modern-day Noah's Ark." Accommodations will be five-star, and membership is by invitation only, to be extended to the world's "elite" families, which are each allotted five thousand square feet of living space to build out to their wildest dreams.

The Russians have an animal-DNA vault. There's a Norwegian seed vault buried in the Arctic. But nobody has built a DNA ark for humans. Why does that seem crazy? What if we're all radiated by something today, tomorrow? The shelter we're planning in Germany is big enough to store the DNA of everyone on the planet. It's the fountain of youth! Soon you'll be able to clone yourself and upload your memory for the future.

While I'm in L.A., Vivos makes public its plans for Europa One. The story gets picked up by blogs such as *ForbesLife* and the *Drudge Report.* In three days, the article on the *Forbes* site is viewed half a million times.

Building a shelter replaces dread with a to-do list, while at the same time fanning the flames of that fear, which offers its own kind of fun, as we invent elaborate and exciting scenarios of our own demise. We embrace the death wish. Safe *in here,* everything *out there* is cause for concern. Drug-resistant super bugs, West Nile, Lyme, Ebola, cancer, bird flu, swine flu, solar storms, climate change, drought, plague, pole shifts, political upheaval, overpopulation, terrorism, supervolcanoes, earthquakes, tornados, tsunamis, economic meltdown, electromagnetic pulses, nuclear war, killer asteroids, the death of the bees, the rise of the machines, the coming of the Rapture, the zombie apocalypse.

Fantasizing about the end is a way of sidestepping responsibility—of overlooking the problems, or systems of inequality, that we might actually be contributing to. (Say, disparities in class and race, global inequality, our environmental footprint, to name just a few.) Destruction always finds the right door to knock on. Our fate is even sealed by science, which tells us we inhabit a galaxy spinning outward into nothing, where by various immutable laws of physics (entropy, thermodynamics), we wind up on the other end of eternity hopelessly cold and alone.

In 1950, atomic scientist Enrico Fermi put a question to Edward Teller, the father of the hydrogen bomb: given the vastness of the universe, and our own unremarkable place in it, why haven't we heard from another civilization?

In other words, *where the hell is everybody?*

This disconnect—infinite universe, no sign of life—is known as the Fermi paradox. It points to the unsettling underbelly of progress. Science advances, but we don't necessarily become more enlightened. Technology tends toward the dark side. There might be a pattern borne out across the universe: civilizations develop to the point where they have the power to do themselves in—and then pull the trigger. The cosmos could be little more than a celestial graveyard, each dead planet just someone else's failed start.

My mind filled with bunkers, I seek higher ground. To the Griffith Observatory on the top of Mount Hollywood—an Art Deco masterpiece that has brought the public closer to the stars since 1935. (Built to withstand an earthquake, the white concrete temple with its three copper domes sits near the Hollywood sign and is itself a screen icon, having been featured in, among others, *Rebel Without a Cause* and three *Terminator* movies.)

A forty-foot-long pendulum swings in the central rotunda—the 240-pound brass ball stays steady in its course, while the earth spins beneath it. A display talks of bursts of gamma rays coming from distant collisions in space. Another discusses the asteroid that ended the dinosaurs. People are eating in the Café at the End of the Universe. They hardly seem worried. From the roof of the observatory, joggers dot the trails winding through the canyons. The city stretches before me in a gray haze. I think of smoke and the wildfires that ravage these hills. I picture the houses on stilts buckling in a mudslide. The sun will be setting soon. There's much I cannot see in the smog, but I know it's all out there. The ocean, for instance—the waves endlessly scraping away at the shore—or, closer in, the ancient tar pits, where mastodons, saber-toothed cats, sloths, and the like—a collection of creatures whose numbers came up a long time ago—trapped themselves unwittingly and sank underground, where they waited, entombed in asphalt, for someone else to come along and dig up the bones.

Everyone asks me when—when is it going to happen? So what if I gave you a date? An event? Are you going to heed the call? Nature has a cleansing process. The wise and strong—they will survive. You know "Vivos" means "to live" in Latin, don't you?

Does an international network of hardened bunkers truly exist? In some ways it doesn't matter. Vicino has tapped a vein. The story has legs. Today's apocalyptic imagination runs wild. During our waking moments—going to work, commuting home, taking in the dreary news of the day, the old existential treadmill—we dream of someone, somewhere, who knows the score, who's anticipating the spectacular. We lean toward that person, hoping to infuse our confused lives with meaning. Part of us wants the train of history to jump its rickety tracks—it's a brave, easy thrill. So we come to inhabit these bunkers of the mind, a string of shelters running beneath our feet, everywhere and nowhere, hidden just out of sight.

The elite have multiple shelters. I know. I've talked to the head of security for Bill Gates; he told me, "He's got them everywhere." *They have theirs. Do you have* yours? *The government has been building big bunkers deep underground. They've got a high-speed subway, or railway, with hubs across the country. They say you can go from D.C. to L.A. in twenty minutes. In the 1990s, small towns all over America reported the sounds of locomotives underground. That was them building it. Just search online for "Air Force underground tunnel machine." I didn't know the Air Force flew underground! This thing is nuclear powered and glazes the rock, so it doesn't collapse.*

I take the bait, and, indeed, a website points to a patent filed in 1971 by Los Alamos scientists, "Method and Apparatus for Tunneling by Melting."

"Are you ready for more?" Vicino asks me again and again.

At least one shelter does indeed stand—in Indiana, where some Vivos members waited out the so-called "Mayan apocalypse" in late December 2012, before eventually going home (relieved? disappointed?) for the holidays. I watch a video tour. The bunker looks nice. Thirty feet down and behind a twelve-inch blast door, there is room for eighty people, plus a minifarm (with grow lights), a medical center, a workshop, a kitchen (granite counters, checkered tile backsplash), a lounge, a theater, a dining room, and a kennel. Vicino calls it an underground cruise ship. It is another retrofit—a Cold War relic repurposed for the future. Vivos is currently taking applications—spots are $35,000 for an adult, $25,000 for a child—but space is limited.

What goes around comes around. When you say "the future," you're really saying "the past." All events are cyclical. Our future has happened many times before. We live in an endlessly changing world. The problem is getting the word out. People think, change is not going to happen in my lifetime. We make predictions, but they don't listen; they're closed-minded. I tell them, "You're keeping your head in the sand, but your ass is hanging out."

What can it all mean? How do we read the signs? Is history interconnected, a series of events that add up to a meaningful pattern that is somehow not of our own making? Or, looking back, do we only see our own reflection? Is that the true final tragedy—to be trapped in our own private visions of history? My interest in the past has always been how it intrudes into the present—which then might suggest a course for the future. But what if that fundamental understanding is flawed? What if the arrow doesn't only point in one direction?

I clip an article from a British newspaper about a strange physics experiment that shows influence runs forward *and* backward in time—at least on a quantum scale. The experiment is complex and involves tiny particles, covert systems, hidden results, and equations that run backward in order to make "retrodictions." The conclusion is startlingly clear: the future can change the past.

Expanded to the human scale, this finding suggests where we are today might be informed retroactively by where we will end up. I don't know what to make of this, what it says about us now. Look around and what do you see—signs of salvation or doom? Put another way: is this bunker building a wake-up call or just the final straw?

Nine months later, I find myself back in L.A. The world hasn't ended yet. I'm staying downtown for a conference. The homelessness is astonishing—particularly along skid row, which occupies four square miles—but also in the business district, which on a deserted early weekend morning looks positively postapocalyptic: empty skyscrapers looming above while a few tourists searching for breakfast weave purposefully between the slow-moving homeless, many of whom are hungry, hopeless, mentally ill. A man shuffles across the street only to park himself in a corner and stare at the wall. The scene is heartbreaking. A friend says it reminds him of *Fear the Walking Dead*, which is set in L.A.—but that's just another way to dehumanize, to duck responsibility, to ignore the debt we owe one another.

I crawl across the city through dirty, gridlocked freeways—the 5, the 101—lines of cars belching fumes to the gray sky. In front, a pickup carries old tires; behind, a graffiti-tagged truck begs for NO WAR. It's overcast and slightly chilly—a period of surprising "June gloom" in this city of the eternally bright seventy degrees. There is a central incongruity to thinking about death in L.A.—an ugliness beneath the radiant veneer, something reckless and sinister. This notion is nothing new—others have put their fingers on this dark pulse—but still I'm surprised to find myself feeling this way as the clouds clear, traffic thins, and I start speeding through a sun-kissed land of swimming pools and fruit trees and beautiful people living nervously—but still living, still living!—under endless blue skies.

About nine miles from downtown, an industrial complex rises beside the freeway. Rusty warehouses straddle both sides of the street—hammers crashing, the cutting of steel. Inside, guys use hooks to lift heavy sheets of metal. Outside, bomb shelters are stacked to the end of the block. They're made of giant corrugated pipes with tubes sprouting from the ends. A thick metal door lies on the ground next to some black vents and long, heavy elbow joints that will become hidden escape tunnels. An unattached blast hatch flips open hydraulically—one day it will reveal a ladder leading down into a mudroom-slash-decontamination shower, where survivors will turn ninety degrees to enter a bunker, thus thwarting any gamma radiation traveling down the hatch.

Having given up on Vivos, I have contacted another company that sells more traditional backyard bunkers.

At last, I have found a shelter ready to take me in.

But only after Ron Hubbard—president and CEO of Atlas Survival Shelters—comes out to greet me. After he asks me not to take pictures. ("We have all you need on the website.") After he tells me he makes all of this stuff except the air systems, which are Swiss or Israeli. ("I'm a world-class steel fabricator. I've won awards.") After he tells me his main business is in ornamental iron—and shows me heavy wrought-iron doors stacked high in a warehouse. After a feral black cat springs from a pile of metal and crosses our path. After he tells me he could sell more than the four bunkers he's able to build each month, but he only breaks even on them, so why rush to make more? ("I'm just here to help people. Those other guys are fearmongers trying to get into your bank account.") After he tells me he ships his bunkers mainly to "Texas, Utah, Indiana . . . everywhere. We'd sell more in California if people had land." After he gives me a glossy brochure. After he brags, "I've been on TV fifty times—I've been on every goddang network there is!" After he leads me into his office, the walls of which are covered with articles (on such topics as how to evade high-altitude electromagnetic-pulse attacks), plus blown-up photos and blueprints of bunkers—many showing his competitors' designs, the flaws of which Ron walks me through carefully. ("It's a small community. Everyone knows what everyone else is doing.") After he tells me how he bought the rights to the name Atlas Bomb Shelter and pulls out an old company brochure from the Cold War certified by the Office of Civil Defense. After he tells me that the corrugated pipe design is the only one that has been tested by the government with an actual nuclear bomb. ("It's the only one they have statistics on.") After he assures me the shockwave will travel through the top six feet of earth, so he buries his bunkers at least ten feet, which has a side benefit of keeping the main hatch a relatively constant fifty-eight degrees. After he shows me pictures of him going down a decommissioned nuclear-missile silo in Nebraska. ("See—there's the corrugated pipe!") After he says, "Everything I do is from a government design. I studied these things for years. Then I just brought the old to the new." After he lists several of his innovations: that angular entry designed to thwart gamma rays, a buried generator pod ("So your enemies can't steal it!"), tamper-proof air filters (plus a backup CO_2 scrubber in case unfriendlies try to block your air pipes), and a secret escape hatch ("If someone comes after my bunker, I can flank 'em through my tunnel and shoot 'em in the back!"). After he promises, "The government contractors said I'm doing it right."

Then, and only then, will Ron lead me to the bunker he will be shipping on Monday. (Next up: a massive doublewide model, some pieces of which are already on the ground.) The bunker before me—which costs $85,900, plus delivery and installation—is ten feet wide and fifty-one feet long, with wood floors and a queen-size bed in the master bedroom, plus eight bunks in another room. There's a kitchen (with space for a full-size fridge), a bathroom (with sink, shower, and toilet), and a living area with a desk, couch, entertainment center, and four-person dinner table. When activated, the air system will make a quiet, comforting drone.

Ron walks to where the bunker sits aboveground on the asphalt—a bright metal shell, careless and cold. Someone else's secret nightmare, waiting to be buried. "Hop up," Ron says, and I jump into the open pipe. I'm standing in the decontamination room. In front of me, the blast door is shut. "Spin the wheel," Ron says. One arrow points up, another points down. Which one will it be? "Down," Ron says. The wheel turns, then stops with a satisfying *thunk*. Ron stays behind. The door swings open and I step into the bunker, alone and unprepared—uncertain how I got here, even less sure where I might be headed.

2015–2016

Bibliography

In writing these essays, I consulted countless articles, maps, brochures, pamphlets, photographs, archival materials, essays, blog posts, and songs. I am indebted to all of my sources, but the following are the books that I leaned on the most.

Alexievich, Svetlana. *Voices from Chernobyl: The Oral History of a Nuclear Disaster.*

Ascher, Kate. *The Works: Anatomy of a City.*

Birk, Dorothy Daniels. *The World Came to St. Louis: A Visit to the 1904 World's Fair.*

Bone, Kevin, and Gina Pollara, eds. *Water-Works: The Architecture and Engineering of the New York City Water Supply.*

Clevenger, Martha R., ed. *"Indescribably Grand": Diaries and Letters from the 1904 World's Fair.*

Compton, Arthur Holly. *Atomic Quest: A Personal Narrative.*

The Cosmopolitan, September 1904.

Court, Darren. *White Sands Missile Range.*

Daily Official Program: World's Fair, St. Louis, June 16, 1904.

Dealey, Ted. *Diaper Days of Dallas.*

Didion, Joan. *Slouching Towards Bethlehem.*

———. *The White Album.*

Early, Gerald, ed. *"Ain't But a Place": An Anthology of African American Writings about St. Louis.*

Federal Writers' Project of the Works Progress Administration for the State of North Dakota. *North Dakota: A Guide to the Northern Prairie State.*

Gerstell, Richard. *How to Survive an Atomic Bomb.*

Greenberg, Stanley. *Invisible New York: The Hidden Infrastructure of the City.*

Greene, A. C. *A Place Called Dallas.*

Hazel, Michael V. *Dallas: A History of "Big D."*

Hill, Patricia Evridge. *Dallas: The Making of a Modern City.*

Hodge, Nathan, and Sharon Weinberger. *A Nuclear Family Vacation: Travels in the World of Atomic Weaponry.*

Horner, Jack. *Dinosaurs under the Big Sky.*

Kolbert, Elizabeth. *The Sixth Extinction: An Unnatural History.*

Kuran, Peter. *How to Photograph an Atomic Bomb.*

Los Alamos Scientific Laboratory Public Relations. *Los Alamos: Beginning of an Era, 1943–1945.*

Loughlin, Caroline, and Catherine Anderson. *Forest Park.*

Manning, Phillip Lars. *Grave Secrets of Dinosaurs: Soft Tissues and Hard Science.*

———. *Dinomummy: The Life, Death, and Discovery of Dakota, a Dinosaur from Hell Creek.*

McMurtry, Larry. *In a Narrow Grave: Essays on Texas.*

Minkin, Bert. *Legacies of the St. Louis World's Fair.*

Nelson, Craig. *The Age of Radiance: The Epic Rise and Dramatic Fall of the Atomic Era.*

Norris, Kathleen. *Dakota: A Spiritual Geography.*

Official Guide to the Louisiana Purchase Exposition, 1904.

Rademacher, Diane. *Still Shining! Discovering Lost Treasures from the 1904 St. Louis World's Fair.*

Rhodes, Richard. *The Making of the Atomic Bomb.*

Rogers, John William. *The Lusty Texans of Dallas.*

Savage, Charles C. *Architecture of the Private Streets of St. Louis: The Architects and the Houses They Designed.*

Schlosser, Eric. *Command and Control: Nuclear Weapons, the Damascus Incident, and the Illusion of Safety.*

Schmitt, Harrison H. *Return to the Moon: Exploration, Enterprise, and Energy in the Human Settlement of Space.*

Sonderman, Joe, and Mike Truax. *St. Louis: The 1904 World's Fair.*

Steinbeck, John. *Travels with Charley: In Search of America.*

Stevens, C. M. *Uncle Jeremiah and His Neighbors at the St. Louis Exposition.*

Szasz, Ferenc Morton. *The Day the Sun Rose Twice: The Story of the Trinity Site Nuclear Explosion, July 16, 1945.*

Vanderbilt, Tom. *Survival City: Adventures among the Ruins of Atomic America.*

Vick, Frances Brannen, ed. *Literary Dallas.*

Witherspoon, Margaret Johanson. *Remembering the St. Louis World's Fair.*

White, E. B. *Here is New York.*

Acknowledgments

This book would not have been possible without the help of many people who generously gave their time, support, and expertise. The book is far better for their efforts. The mistakes remain all mine.

This is but a partial list of those to whom I am grateful:

For assistance with the various essays: Taylor Hamra (third book in a row), Lorin Stein, Sadie Stein, Shawn Rosenheim, Bruce Falconer, Dan Piepenbring, Gerald Early, Saaba Buddenhagen Lutzeler, Joshua Gang, the Missouri History Museum Library and Research Center, Yuka Igarashi, Todd Rohal, Charles Gandee, Pilar Viladas, the staff of *Talk* magazine, Harrison H. Schmitt, Teresa A. Fitzgibbon, Rowena Baca, Robb Hermes, Bradley L. Jolliff, the Washington University Libraries' Department of Special Collections, Tyler Lyson, Kirk Johnson, Antoine Bercovici, Ranse and Molly Lyson, Tom Tucker, Stephen Begin, Stephen Floersheimer, Barb Benty, Suzy McIntire, the staff of the Marmarth Research Foundation, Barbi Grossman, Robert Vicino, Ron Hubbard, and Micah Levin.

For their professional advice: Jeff Shotts, W. Ralph Eubanks, Brigid Hughes, Cheston Knapp, and Andrew Leland.

My brilliant university colleagues, including but not limited to: Mary Jo Bang, Kathryn Davis, Danielle Dutton, Kathleen Finneran, Marshall Klimasewiski, Carl Phillips, David Schuman, Barbara Schaal, Wolfram Schmidgen, Vincent Sherry, and the Wash. U. English Department. Plus Saher Alam and Devin Johnston at St. Louis University.

The incomparable team at Coffee House Press—for their dedication to books and their intelligence with this one—particularly Chris Fischbach, Caroline Casey, Carla Valadez, Amelia Foster, Mandy Medley, Nica Carrillo, Lizzie Davis, and Erika Stevens. Plus Maya Beck, Annemarie Eayrs, and Blake Planty for checking the facts.

My tireless and amazing agent, Emma Parry, and the rest of the team at Janklow & Nesbit, particularly Chad Luibl.

My friends and teachers: Jim Shepard, Charles Baxter, Patricia Hampl, Julie Schumacher, John Irving, Rebecca Traister, Stacy Cochran, Tod Lippy, and Ethan Rutherford.

My family: Ted and Sally McPherson, Bud and Sherry Larson, and Beth and Josh Gaffga.

Most of all, I thank Heather, my first reader, for her editorial ruthlessness and—along with Penny—for suffering my absences and preoccupations. You two remain the reason I think about these things.

Coffee House Press began as a small letterpress operation in 1972 and has grown into an internationally renowned nonprofit publisher of literary fiction, essay, poetry, and other work that doesn't fit neatly into genre categories.

Coffee House is both a publisher and an arts organization. Through our *Books in Action* program and publications, we've become interdisciplinary collaborators and incubators for new work and audience experiences. Our vision for the future is one where a publisher is a catalyst and connector.

LITERATURE
is not the same thing as
PUBLISHING

Funder Acknowledgments

Coffee House Press is an internationally renowned independent book publisher and arts nonprofit based in Minneapolis, MN; through its literary publications and *Books in Action* program, Coffee House acts as a catalyst and connector—between authors and readers, ideas and resources, creativity and community, inspiration and action.

Coffee House Press books are made possible through the generous support of grants and donations from corporations, state and federal grant programs, family foundations, and the many individuals who believe in the transformational power of literature. This activity is made possible by the voters of Minnesota through a Minnesota State Arts Board Operating Support grant, thanks to the legislative appropriation from the arts and cultural heritage fund. Coffee House also receives major operating support from the Amazon Literary Partnership, the Bush Foundation, the Jerome Foundation, The McKnight Foundation, Target Foundation, and the National Endowment for the Arts (NEA). To find out more about how NEA grants impact individuals and communities, visit www.arts.gov.

Coffee House Press receives additional support from the Elmer L. & Eleanor J. Andersen Foundation; the David & Mary Anderson Family Foundation; the Buuck Family Foundation; the Carolyn Foundation; the Dorsey & Whitney Foundation; Dorsey & Whitney LLP; the E. Thomas Binger and Rebecca Rand Fund of the Minneapolis Foundation, the Knight Foundation; the Rehael Fund of the Minneapolis Foundation; the Matching Grant Program Fund of the Minneapolis Foundation; the Schwab Charitable Fund; Schwegman, Lundberg & Woessner, P.A.; the Scott Family Foundation; the US Bank Foundation; VSA Minnesota for the Metropolitan Regional Arts Council; the Archie D. & Bertha H. Walker Foundation; and the Woessner Freeman Family Foundation in honor of Allan Kornblum.

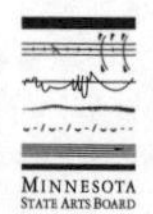

The Publisher's Circle of Coffee House Press

Publisher's Circle members make significant contributions to Coffee House Press's annual giving campaign. Understanding that a strong financial base is necessary for the press to meet the challenges and opportunities that arise each year, this group plays a crucial part in the success of Coffee House's mission.

Recent Publisher's Circle members include many anonymous donors, Mr. & Mrs. Rand L. Alexander, Suzanne Allen, Patricia A. Beithon, Bill Berkson & Connie Lewallen, Robert & Gail Buuck, Claire Casey, Louise Copeland, Jane Dalrymple-Hollo, Ruth Stricker Dayton, Jennifer Kwon Dobbs & Stefan Liess, Mary Ebert & Paul Stembler, Chris Fischbach & Katie Dublinski, Kaywin Feldman & Jim Lutz, Sally French, Jocelyn Hale & Glenn Miller, the Rehael Fund-Roger Hale/Nor Hall of the Minneapolis Foundation, Randy Hartten & Ron Lotz, Jeffrey Hom, Carl & Heidi Horsch, Amy L. Hubbard & Geoffrey J. Kehoe Fund, Kenneth Kahn & Susan Dicker, Stephen & Isabel Keating, Kenneth Koch Literary Estate, Jennifer Komar & Enrique Olivarez, Allan & Cinda Kornblum, Leslie Larson Maheras, Lenfestey Family Foundation, Sarah Lutman & Rob Rudolph, the Carol & Aaron Mack Charitable Fund of the Minneapolis Foundation, George & Olga Mack, Joshua Mack, Gillian McCain, Mary & Malcolm McDermid, Sjur Midness & Briar Andresen, Maureen Millea Smith & Daniel Smith, Peter Nelson & Jennifer Swenson, Marc Porter & James Hennessy, Robin Preble, Jeffrey Scherer, Jeffrey Sugerman & Sarah Schultz, Nan G. & Stephen C. Swid, Patricia Tilton, Stu Wilson & Melissa Barker, Warren D. Woessner & Iris C. Freeman, Margaret Wurtele, Joanne Von Blon, and Wayne P. Zink.

For more information about the Publisher's Circle and other ways to support Coffee House Press books, authors, and activities, please visit www.coffeehousepress.org/support or contact us at info@coffeehousepress.org.

EDWARD MCPHERSON is the author of two previous books: *Buster Keaton: Tempest in a Flat Hat* (Faber & Faber) and *The Backwash Squeeze and Other Improbable Feats* (HarperCollins). He has written for the *New York Times Magazine*, the *Paris Review*, *Tin House*, and the *American Scholar*, among others. He has received a Pushcart Prize, the Gulf Coast Prize in Fiction, a Minnesota State Arts Board grant, and the Gesell Award from the University of Minnesota, where he received his MFA. He teaches creative writing at Washington University in St. Louis.

The History of the Future was designed by
Bookmobile Design & Publisher Services.
Text is set in Sabon Next LT Pro.